1981

TOPICS

IN

CLASSICAL

BIOPHYSICS

TOPICS

IN

CLASSICAL

BIOPHYSICS

HAROLD J. METCALF

Department of Physics
State University of New York at Stony Brook

PRENTICE-HALL, INC.
Englewood Cliffs, New Jersey   07632

Library of Congress Cataloging in Publication Data

METCALF, HAROLD J (date)
  Topics in classical biophysics

  Bibliography: p.
  Includes index.
  1. Biological physics. I. Title.
QH505.M47 1979   574.1′91   79–10383
ISBN 0–13–925255–X

© 1980 by PRENTICE-HALL, INC.
Englewood Cliffs, New Jersey 07632

Printed in the United States of America

10 9 8 7 6 5 4 3 2 1

Editorial/production supervision
and interior design by Ted Pastrick and Guy Lento
Cover design by Frederick Charles, Ltd.
Manufacturing buyer: John Hall

PRENTICE-HALL INTERNATIONAL, INC., *London*
PRENTICE-HALL OF AUSTRALIA PTY. LIMITED, *Sydney*
PRENTICE-HALL OF CANADA, LTD., *Toronto*
PRENTICE-HALL OF INDIA PRIVATE LIMITED, *New Delhi*
PRENTICE-HALL OF JAPAN, INC., *Tokyo*
PRENTICE-HALL OF SOUTHEAST ASIA PTE. LTD., *Singapore*
WHITEHALL BOOKS LIMITED, *Wellington, New Zealand*

To David, Cindy, and Jonathan

Contents

Preface

The workings of the human body have attracted and fascinated the great minds of every age. In modern times we are witnessing a major step forward in the understanding of problems in the biological sciences through the application of the analytic techniques of physics and chemistry. The use of physics in biology and medicine is not new, but its presence has been essential for the current advances in modern biology and human physiology.

This book is intended for those with some background in physics who wish to learn about its application to biological problems. The reader should have taken a one-year introductory college level course in calculus-based physics. Courses in chemistry, biology, advanced calculus, and advanced physics will enable greater depth of understanding, but are definitely not required for comprehension of the material.

Anyone with an interest in the life sciences should be attracted to this book. Medical (or premedical) students will find that many of the ideas from their required one-year physics courses are discussed in some detail here. Biology and physiology students will find a review of some physical principles and a summary of certain applications of them. Physics and engineering students will find an introduction to some biological phenomena which should encourage their pursuit of further study and analysis. Many students who will be seeking careers in bio-engineering, medical engineering, medical electronics, biophysics, space medicine, and other allied health sciences will find this book attractive.

Furthermore, the book is not intended for students alone. The material has been chosen and presented in a manner compatible with the idea that it may be read by mature engineers or scientists who are contemplating a career change. These people should find that this book introduces several basic physiological phenomena in a language they can understand and in terms familiar to them.

The first chapter is based on simple mechanics. The ideas of ballistic motion, forces and equilibrium, velocity and acceleration, and several others are used to derive simple and interesting information about animal motion. It is shown why lower back injuries are to be expected from improper lifting, why small animals are not injured in falls, and why most animals can run at approximately the same top speed (within a factor of 2 of 10 m/sec) even though their masses vary over several decades.

The second chapter is based on the same fundamental classical mechanics, but the emphasis is on energy and its transport. The metabolic sources of energy are discussed along with some description of heat transfer by various methods. For example, the radiative heat loss from a human body at 310°K is nearly 10,000 Cal/day if there is no return of heat to that body from the surrounding heat bath at 293°K. The ubiquitous phenomenon of counter-current exchange is treated in some detail with a discussion of water conservation by desert animals, heat conservation by arctic mammals, and urine concentration in kidneys. Athletic performance, especially running, is analyzed and the expectations are calculated: Table 2.2 compares the results with Olympic records and the comparison is impressive.

Chapter 3 contains a study of fluids. It starts with hydrostatics and its application to elastic blood vessels. A discussion of surface tension is followed by the beginning of hydrodynamics. The flow of frictionless (Bernoulli) fluids is discussed and applications such as catheterization and veinous collapse are presented. Then a discussion of viscous fluids provides the basis for fluid friction and peripheral circulation control.

Chapter 4 is devoted to blood circulation. The first part describes the plumbing of the circulatory system and the structure of the heart as a recip-rocating pump. Ballistocardiography and electrocardiography are discussed qualitatively. The chapter ends with a discussion of hemodynamics in large, intermediate, and small blood vessels.

The fifth chapter is devoted to feedback and control. The general prop-erties of feedback stabilized systems are presented in the form of simple numerical examples. A general, proportional control, time delay system is described and its oscillatory and divergent properties are presented, but again the solution to the differential equation is simple (no Laplace transform). A discussion of operational amplifiers is followed by several examples in which the principles described in the early part of the chapter are employed.

Chapter 6 contains a classical description of axons. The cable theory of excitable axons is discussed with many numerical calculations. The famous Hodgkin-Huxley equations are presented and their origin is described, but there is no quantitative study of the dynamical behavior of axons. Communication by chemical transmitters is presented with a discussion of the speed and $\sqrt{N}$ character of diffusive processes.

The seventh and eighth chapters are descriptions of sensory perception (of sound and light). Each begins with an introduction to the physical phenomena (waves) and is then followed by anatomical and quantitative descriptions of our sensory organs (ear and eye). Then there are sections which describe the phenomena of perception (psychophysics) and explain a variety of experiences. There are discussions of the role of lateral inhibition in tonal discrimination and image processing. Chapter seven ends with a discussion of music and chapter eight ends with a discussion of painting and color perception.

The last chapter discusses several of the laboratory methods that are used to acquire much of the information that contributes to our knowledge of biology. It begins with a discussion of optical spectroscopy as applied to molecules of biological interest. That is followed by a section on resonance (ESR, NMR, etc.). The general characteristics of a simple magnifying glass leads to a discussion of microscopy and an explanation of various types of modern microscopes and their uses. There is a section on electronic instruments for extracting signals from noise, on light scattering and centrifugation from suspensions, and on ultrasonic diagnostic techniques.

There are several appendices which contain some calculations as well as some descriptions of peripheral material (such as the types of available photodetectors). Each chapter is followed by some problems and questions. Some of these can be answered by a brief sentence or calculation, others are intended to lead to stimulating discussions. The bibliography of each chapter is annotated with descriptive comments about each book or article.

The presentations are not consistent from chapter to chapter because of the diversity of the material. Some parts of the text are very descriptive with little calculation while other parts are quite analytic. In some cases a reader with little background can digest most of the material in a chapter whereas in other cases each point depends upon understanding the previous calculations.

This book was developed from the lecture notes of a third-semester physics course for which the author could find no suitable text. It is not intended to be used as a sole text, but should be supplemented with Scientific American offprints and readings from physiology books. Conversely, it would be fine supplementary reading for a physiology or biophysics course.

This book is not intended to be complete. I have deliberately avoided discussion of two of the most active topics in biophysics: membranes and molecular biophysics. Furthermore, there are many classical ideas which have also been omitted, either deliberately or out of ignorance. I hope that those topics I have chosen will be of interest to the reader.

Acknowledgments

This book is the result of my teaching a course for several years to generations of students whose questions and curiosity provided the stimulus for much of its contents. Their suggestions and contributions led to many of the ideas presented here and I am thankful to them for all that they have taught me.

I am indebted to L. David Roper of The Virginia Polytechnical Institute and to Ted Ducas of Wellesley College for their careful reading of the manuscript and their many helpful suggestions and comments. Thanks also go to many physicians, nurses, medical technicians, biologists, and physiologists who have answered my questions and explained many procedures and devices to me.

I wish to thank Pat Peiliker for her careful typing and patient proof reading. She devoted many long hours to corrections, revisions, indexing and other aspects of the writing of this book. I am grateful to her for her help and contributions. I also wish to thank the staff of Prentice Hall, Inc., especially Logan Campbell and Guy Lento, for their help and advice.

TOPICS
IN
CLASSICAL
BIOPHYSICS

1

BIOMECHANICS

Part 1: Ballistic Motion

Although the study of mechanics dates back to the ancient Egyptians and Greeks, we shall be concerned with modern mechanics that was developed from the seventeenth century work of Galileo and Newton. Biomechanics is the application of their ideas and concepts to the equilibrium and movements of animals. Although the primary topics of this book will be associated with the human body, we shall begin with a study of the motion of various animals and comparisons among them.

Perhaps the most easily studied form of animal motion is free fall, i.e., motion under circumstances where there are no forces on the animal other than gravity. In order to describe this motion we make the approximation of neglecting all forces from the air such as lift generated by wings and air resistance. This approximation pretty well restricts us to the consideration of the jumping or falling motion of animals whose mass is larger than about 100 grams.

We shall regard the gravitational acceleration as constant ($g = 9.8$ m/sec²) and downward. This means that horizontal motion occurs at constant velocity v_x but vertical motion is constantly accelerated. We write

$$v_x = v_{0x} \qquad v_y = v_{0y} + at = v_{0y} - gt \qquad (1.1)$$

where v_y is the vertical component of velocity and v_{0y} is the value of v_y when $t = 0$.

If we wish to calculate the position of an object at any time, we must compute the horizontal and vertical coordinates separately. We integrate Eq. 1.1 and find

$$x = x_0 + v_x t \tag{1.2}$$

for the horizontal coordinate noting that x_0 (position at $t = 0$) and v_x (horizontal velocity) are both constant. In order to compute y we must account for the fact that v_y is not a constant. Since v_y is defined to be dy/dt, we write

$$y = \int v_y \, dt = v_{0y} t - \frac{gt^2}{2} + y_0 \tag{1.3}$$

where y_0 (the constant of integration) is the value of y at $t = 0$.

Equations 1.1 and 1.3 give the velocity and position of a uniformly accelerated object as a function of time. They may be combined to give the vertical velocity as a function of position

$$v^2 = 2g(y_0 - y) + v_0^2 \tag{1.4}$$

where the subscript y has been omitted for clarity. Equation 1.4 can also be derived using conservation of mechanical energy:

$$\tfrac{1}{2}mv_0^2 + mgy_0 = \tfrac{1}{2}mv^2 + mgy. \tag{1.5}$$

There are several special cases to which we apply Eq. 1.4. We note that if an object is projected upward from the ground ($y_0 = 0$) with an initial velocity v_0, it will reach its maximum height when $v = 0$. This height is

$$y = \frac{v_0^2}{2g}. \tag{1.6}$$

We also note that if an object begins to fall from rest ($v_0 = 0$), its velocity is given by

$$v = \sqrt{2gh} \tag{1.7}$$

where $h = y_0 - y$ is the height through which it has fallen.

It is important to remember that if the mass whose motion we are studying is not a point mass, but is instead a jumping animal, the formula above describes only the motion of the center of mass and is *NOT* necessarily correct for any part of the animal. In the following discussion the coordinates refer *only* to motion of the center of mass.

A. If an animal is at rest on the ground and then jumps to a height h, it must accelerate upward to a vertical velocity $v = \sqrt{2gh}$ and achieve a kinetic energy $mv^2/2$ by applying an initial force F on the ground. In order to calculate this applied force we must make certain assumptions about the way the animal produces it. The simplest assumption is that the applied force is constant. (Although this is not at all a realistic description of muscular forces, a more accurate calculation in Appendix A supports the use of the approximation.) If the force is constant, the amount of work done by it on the animal's body is Fs where s is the distance traveled by the animal during the upward acceleration. We use Eq. 1.6 with the condition

$$\tfrac{1}{2}mv_0^2 = Fs \tag{1.8}$$

and find

$$h = \frac{F}{W}s \tag{1.9}$$

where $W = mg$ is the weight of the animal. Of course the animal must also exert the force required to support his own weight at the same time so that his applied force is $W(1 + h/s)$.

We now estimate the quantity s. This is the distance traveled during acceleration time when the animal's feet must be on the ground. Clearly s cannot be larger than the size of the animal and, in fact, must be somewhat smaller than that. For a man jumping from a deep crouch, s is approximately one-third of his height. Similarly, for most other animals, s is between one-third and one-half of their height. Therefore an animal can jump (i.e., raise his center of gravity) to his own height only if he can lift two or three times his weight. Most humans can just about lift their own weight and can just barely jump to one-third of their height. (You can test this by jumping up to reach the highest possible point and seeing that it is barely $\tfrac{1}{2}$ m above the highest point you can reach while standing.) Since the center of gravity of a human is just below the navel and therefore about 1 meter above the ground, this implies a high jump over a bar at about 1.5 m (or about 5 ft). A grasshopper or flea, which can jump many times its own height, obviously has the strength to lift many times its weight. Note that very large animals (elephants) can only jump a small fraction of their height: most of their muscular effort is required just to support their weight and this leaves very little reserve for jumping.

B.   We now consider another application of the ballistics to animal motion. When salmon swim upstream to spawn, they often encounter waterfalls and are frequently seen jumping out of the water and over the falls. In order to jump a waterfall of height h, the fish must achieve a vertical velocity of at least $\sqrt{2gh}$. In addition, the fish must also achieve a horizontal component of velocity relative to the water that is flowing downstream below the falls. This velocity is the sum of two terms, one arising from downstream flow and the other arising from the requirement to provide for some horizontal travel during flight so that the fish will land on top of the falls and not drop back to its starting point. We call the total horizontal component of velocity v_h and find that the net speed $|V|$ required when the fish leaves the water is given by the vector sum of the vertical and horizontal components:

$$|V| = \sqrt{2gh + v_h^2}. \tag{1.10}$$

When the falling water approaches the bottom of the falls, it is traveling downward at the speed $V_v = \sqrt{2gh}$ and if the fish can swim faster than this, he can swim up the falls. In fact, he can do it with a maximum top speed which is considerably less than the V in Eq. 1.10 which is required to jump over the falls. The reason why fish jump over the falls is not clear. It may be, for example, that the fish expends less energy to jump than to swim simply because of the length of time required to swim up the falls.

C. A third application of ballistic motion appears when we try to calculate how far a person can throw something. It is shown in Appendix B that an object thrown at a 45° angle at velocity v will travel a horizontal distance $d = v^2/g$ before landing. We assume uniform applied force F during acceleration from rest to launch velocity (see Appendix A) and find that the kinetic energy imparted to the object is

$$\tfrac{1}{2}mv^2 = Fs \tag{1.11}$$

where s is the distance covered during the acceleration period. The magnitude of F will be estimated from knowing how much weight, $F = Mg$, can be lifted. The distance d is

$$d = \frac{v^2}{g} = \frac{2Fs}{mg} = 2s\frac{M}{m}. \tag{1.12}$$

Let us substitute some reasonable numbers into Eq. 1.12 and calculate d. A baseball player can easily lift a mass of 20 kg, and the baseball he throws has a mass of $\frac{1}{2}$ kg. Assume the length s over which he can accelerate a baseball is approximately 1 meter. (A good thrower steps into the throw in order to make s as large as possible.) Then we find $d = 80$ m (or about 250 ft), which is a typical "good" throw. The speed a player (pitcher) can achieve is found from Eq. 1.12 to be $v = \sqrt{2sMg/m} \cong 28$ m/sec, which is more than 60 mph. This formula gives reasonable results for a baseball and a shotput. (The Olympic record for the $\cong 7$ kg shot is about 18 m but a shot putter is much stronger than a baseball player and can lift 60 kg.)

We have seen some examples of simple motion described quantitatively and it is suggested that the reader try to invent a few more.

Part 2 : Scaling

After our brief study of the mechanics of motion, we now discuss a topic called *scaling*. Consider geometrically similar objects whose linear dimension is characterized by the length l and whose volumes are proportional to but not equal to l^3. We write $V \sim l^3$ and refer to such a statement as an allometric equation. In this section we shall use allometric equations to determine a variety of comparative characteristics of animals. We shall use a number of necessary and plausible assumptions. The most important of these is that all animals we consider are geometrically similar. Even though this does not seem to be a particularly good assumption at the outset, its consequences are consistent with a variety of observations. We also assume that the density of animal tissue is the same for all species and that all species have about the same body temperature. The first of these assumptions leads directly to the allometric statement that animal weights $W \sim l^3$.

A. It is well known (and obvious) that the maximum force that can be exerted by a muscle is proportional to the cross-section area of that muscle. We write $F \sim l^2$. When an animal jumps up, its acceleration determined by this muscular force is

$$a = \frac{F}{m} = g\frac{F}{W} \sim \frac{l^2}{l^3} \sim l^{-1}. \tag{1.13}$$

The launch velocity is therefore given by (see Eq. 1.6)

$$v = \sqrt{2as} \sim \sqrt{l^{-1}l} \sim l^0 \tag{1.14}$$

where s is the vertical distance traveled during the acceleration while the animal's feet are still on the ground. Clearly $s \sim l$ as indicated in Eq. 1.14 and the launch velocity v is *independent* of l. The final height of the jump h is given by $h = v^2/2g$ and is therefore also independent of animal size. The astounding conclusion that the maximum height of animal jumps is independent of size is borne out by the following observation: there are very few animals, from the smallest mouse to the largest elephant, whose maximum jumping height (the distance by which the center of gravity is raised) is not within a factor of 2 of $\frac{2}{3}$ meter (see Figure 1.1). The weight of these animals ranges over a factor of ten thousand, but their jumping height only varies by about a factor of 2.

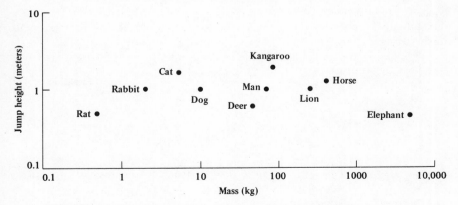

Figure 1.1 Plot of the height to which various animals can jump. Over a range of a factor of 10,000 in weight the jump height varies by little more than a factor of 2.

B. A similar calculation for running speed was given by J. M. Smith. He assumed that when an animal is running, its speed is limited by the power required to overcome the inertia of the limbs. The muscle must shorten by an amount $d \sim l$ where l is the limb length, exert a force F, and provide the limb with an angular velocity $\omega = v/l$ where v is the running speed. In order to take a stride the muscle does an amount of work

$$W = Fd \sim l^2 l \sim l^3. \tag{1.15}$$

On the other hand, the energy imparted to the limb in each stride is proportional to $I\omega^2$ where I, the moment of inertia, is proportional to $ml^2 \sim l^3 l^2 \sim l^5$. Then we find the energy

$$E \sim l^5 \frac{v^2}{l^2} \sim l^3 v^2. \tag{1.16}$$

Since the energy E imparted to the limb must equal the work input W, we see that $E \sim l^3 v^2$ (from Eq. 1.16) and $W \sim l^3$ (from earlier discussion) requires v independent of l.

The same conclusion appears if we assume the primary effort is overcoming air friction. The maximum power output scales with l^2 (see Part 2G on p. 9). If all the power is used to overcome air friction, the power requirement $P = Fv = kAv^3 \sim l^2 v^3$ (see Appendix C on air friction). However, available power $P \sim l^2$ so that the speed v is independent of l. This conclusion is also borne out by observation. The top running speed of most animals is within a factor of 2 of 7 m/sec or about 15 mph (see Figure 1.2). Smith then goes on to say that "much of the energy of the limb is degraded into heat in the antagonistic muscles during the second half of the (running) stroke. This is the price paid by animals for not having wheels, and explains the advantages of riding a bicycle."

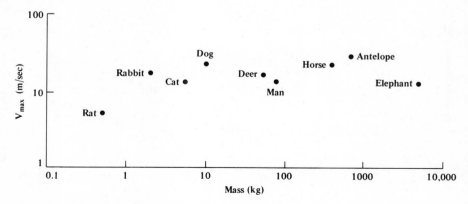

Figure 1.2 Plot of the running speed of various animals. Over a range of a factor of 10,000 in weight, the running speed varies by little more than a factor of 3.

C. Smith continues with the following discussion. The power P required to run is the amount of energy used in each step divided by the time t required for it. Further, the speed v and the time t are related: $t = z/v \sim l$ where z is the stride length and $z \sim l$. Then we have

$$P = \frac{E}{t} \sim \frac{l^3}{l} \sim l^2. \tag{1.17}$$

If an animal is to run uphill, it must do the same amount of work as is required to run on the flat plus an additional amount to raise its body against gravity. The power required to work against gravity P_g is the weight times the vertical velocity and therefore $P_g \sim l^3 v$. Since the power available increases only with l^2, it must be that for uphill running $v \sim l^{-1}$. It is certainly observed that large animals slow down much more than small ones when running uphill.

D. The ideas of scaling are very easily applied to the study of bone structure. Since the bones of an animal must support its weight W, it is clear that their strength should increase in direct proportion to the animal's weight. On the other hand, since the strength of the bone material is assumed to be the same from one animal to another, then the strength of a particular bone is proportional to its cross-section area A, which depends on the square of the radius r. We have

$$A = \pi r^2 \sim W \sim l^3 \tag{1.18}$$

which leads to

$$r \sim l^{3/2}. \tag{1.19}$$

This means that the bones of similarly shaped animals are *not* similar. As the animals become larger, the bones become relatively thicker in accordance with a 3/2 power law. This assertion is qualitatively supported by the data of Kayser and Heusner plotted in Schmidt-Nielsen's book and by observation of animal skeletons.

E. Earlier in this chapter we considered the ballistic motion of point masses. In fact, when bodies fall freely in the earth's gravitational field, they do not accelerate continuously because the friction from the air reduces the acceleration. The friction force F is proportional to the cross-sectional area of the falling body and to the square of the velocity in the range of velocities we consider here (see Appendix C). We must add the velocity-dependent friction force to the weight so that

$$\Sigma F = ma = -mg + kAv^2 \tag{1.20}$$

where k is the friction proportionality constant. Acceleration ceases when velocity has increased so that $mg = kAv^2$ and the value of v for which this occurs is called the *terminal velocity* v_t. We have $v_t^2 = mg/kA \sim l^3/l^2 \sim l$ or $v_t \sim l^{1/2}$. Larger animals have higher terminal velocities.

In order for an animal to land safely when falling at terminal velocity, its muscles must be able to exert a force large enough to decelerate it. If the deceleration force is constant (see Appendix A), the deceleration a is simply $a = v_t^2/2s$ where s is some fraction of body length. The force F required for this deceleration is ma and we have

$$F = \frac{mv_t^2}{2s} \sim \frac{l^3 l}{l} \sim l^3. \tag{1.21}$$

Since the maximum force a muscle can exert depends on l^2, it is clear why larger animals cannot land safely from their terminal velocities but smaller animals can survive a fall from any height.

We can estimate the height from which an animal must fall in order to reach terminal velocity by ignoring air friction for all velocities less than v_t. The height we find is simply $v_t^2/2g$. Parachutists and sky divers tell us that, for a man, v_t is about 100 mph or about 50 m/sec. A man must fall 125 m or nearly 400 ft to achieve this velocity. Since a 125-m fall is nearly always fatal to man, it is clear that men cannot survive a landing at v_t. On the other hand, a 50-g mouse has a linear dimension about one-twelfth that of a man and his terminal velocity is therefore 29 mph or about 14 m/sec. The height from which he falls to reach v_t is 10 m and he may survive such a fall. Any animal smaller than a mouse certainly need not concern itself with the danger of falling. In fact, some insects can jump high enough to land near their terminal velocities.

F.   We now consider applications of scaling to animal function, particularly circulation. In order to determine how pulse rates scale, we assume that one of the primary purposes of blood circulation is to bring body heat from a warm core to the body surface in order to keep surface tissue warm. The heat loss from the surface is, of course, dependent on its area, which increases as l^2. The blood is pumped by a heart whose linear dimension increases with the linear dimension of the animal and whose stroke volume therefore increases with l^3. The total amount of blood pumped in a fixed time therefore depends on $l^3 f$ where f is the pulse rate. In order to balance the heat loss, which varies with l^2, with the heat supply, $l^3 f$, we have $l^2 \sim l^3 f$ or $f \sim l^{-1}$. Therefore larger animals have slower heart rates. This conclusion is not only supported by data from one animal to the next but also by data of heart rate versus size for the same species, as shown in Figure 1.3. The normal heart rate of humans decreases from infancy through childhood into maturity according to $f \sim l^{-1} \sim W^{-1/3}$ where $W \sim l^3$ is the weight. It does not depend in any simple way on age.

It is observed that the average life span of various animals increases roughly as l (see Figure 1.3). This implies that animal hearts all beat approximately the same number of times in their lifetime. This number is about 6×10^8.

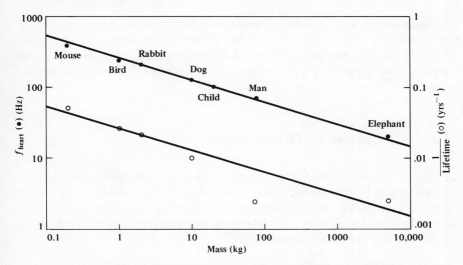

Figure 1.3 Plot of heart rate (reciprocal of time between beats) and death rate (reciprocal of lifetimes) of various animals versus weight. The straight line of the heart-rate curve has a slope of almost exactly $-\frac{1}{3}$ as discussed in the text. The parallelness of the two lines indicates that there is a strong and simple correlation between the rates. Note that man is anomalously long-lived.

G. We notice that a heart rate which scales as l^{-1} can supply heat (either directly or by carrying nutrients) at a rate which scales as l^2. Therefore the amount of power available to an animal scales with l^2. This is consistent with our calculations for muscle output (see earlier discussion on running). Furthermore, the l^2 dependence of energy consumption suggests that the daily food requirement of animals scales with l^2 assuming that the purpose of eating is to supply heat to keep warm and assuming that food has about the same amount of energy per unit weight. Since weight scales with l^3, the fraction of weight that must be eaten daily scales as l^{-1}. A man eats about 1.5 kg of food daily, which is about $\frac{1}{50}$ of his weight. A mouse, which weighs about 2000 times less than a man (l smaller by a factor of about 12), needs $\frac{1}{150}$ as much food or about 10 g. However, this 10 g is about 25 % of his body weight, nearly 2 weeks worth of food for a man.

Much of this discussion has been based on the l^2 dependence of energy requirement that implies an l^2 dependence of metabolism rate. Since we have used $W \sim l^3$, we have implicitly assumed that metabolism scales with $W^{2/3}$. Although the metabolism data are fitted very well by a $W^{2/3}$ law, they are fitted better by $W^{3/4}$. McMahon has presented a detailed discussion of this topic and has arrived at an argument that produces a $W^{3/4}$ scaling for metabolism rate. His reasoning, however, depends on the assumption that the body's limbs are uniform, homogeneous, and self-supporting, which is simply not so. Our limbs are supported by an internal skeleton whose scaling is consistent with his argument, but the shape of the limbs themselves do not necessarily scale the same way that the bones do. The discussions in this section, however, are all independent of whether the correct exponent is $\frac{2}{3}$ or $\frac{3}{4}$.

Part 3 : Forces and Equilibrium

In this section we shall study the mechanical equilibrium of certain structures in the human body in order to learn about the forces, pressures, and strains created by muscles. The first structure we study is the spinal column. We begin by considering it as a straight, rigid rod and will later account for its curvature and vertebrate construction.

A. Consider what happens when a person bends over at the hips so that his (straight) spine of length l makes an angle θ to the horizontal, as shown in Figure 1.4. In order to achieve equilibrium, i.e., not fall over, it is required

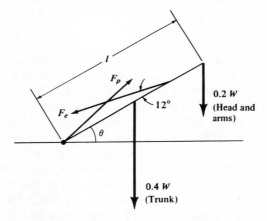

Figure 1.4 Vector diagram of forces on a schematic human torso. The spine is represented as a straight, rigid bar pivoted at the hip. The weights of various body parts are estimated.

that the vector sum of the forces and simultaneously the vector sum of the torques on the spine be zero. There are four forces to consider. First, there is the weight of the arms and head. We assume this force is downward with a magnitude of $0.2W$ where W is the weight of the body. It is applied right at the top end of the spine. Second, we consider that the weight of the trunk is $0.4W$ and that it acts at the center of mass, namely halfway up the spine. Third, there is the force F_p produced at the pelvis and applied at the base of the spine. This force supports the weight of the trunk, head, and arms. All of these forces produce torques that tend to rotate the spine clockwise in the diagram. These torques are balanced by the fourth force F_e from the muscles called *erector spinae*. Anatomical studies show that the force is applied at a point about two-thirds the way along the spine and at an angle of about $12° \cong 0.2$ radian.

We now write down the equilibrium conditions, beginning with the torque. An astute choice of the origin at the point of application of F_p will make things much easier for us by removing this force from the torque equation. We are left with

$$\Sigma T = 0 =$$

$$F_e \times \tfrac{2}{3}l \times \sin 12° - 0.4W \times \tfrac{1}{2}l \times \sin (90 + \theta)° - 0.2Wl \sin (90 + \theta)° \quad (1.22)$$

since $\vec{T} = \vec{r} \times \vec{F}$ and the magnitude of the torque is given by the product of the magnitudes of r and F times the sine of the angle between them. We now solve for F_e and find

$$F_e = \frac{0.6W \sin (90 + \theta)°}{\sin 12°} \cong 3W \cos \theta \quad (1.23)$$

and if we let $\theta = 30°$ and $W = 160$ lb (75 kg), we find the rather large force $F_e = 2.5 \, W = 400$ lb (180 kg). Notice that our choice of origin has made this solution possible because l, the length of the spine, has dropped out.

Next we use the force equilibrium to find F_p. For the vertical and horizontal components we have

$$\Sigma F_y = 0 = -0.4W - 0.2W - F_e \sin (\theta - 12)° + F_p(y) \quad (1.24a)$$

$$\Sigma F_x = 0 = -F_e \cos (\theta - 12)° + F_p(x). \quad (1.24b)$$

We find

$$F_p(y) = W[0.6 + 3 \cos \theta \sin (\theta - 12)°] \quad (1.25a)$$

and

$$F_p(x) = 3W \cos \theta \cos (\theta - 12)° \quad (1.25b)$$

so that the direction of F_p is at an angle ϕ above the horizontal where

$$\tan \phi = \frac{\tfrac{1}{3} \sec \theta + \sin (\theta - 12)°}{\cos (\theta - 12)°}. \quad (1.26)$$

The component of F_p along the spinal column for $\theta = 30°$ and $W = 160$ lb (75 kg) is $2.8W$ or 446 lb (210 kg). This very large force acts directly along the spinal column.

B. We now consider that the spine is not a single rod but is actually constructed of a stack of bones called *vertebrae*, each separated from its neighbor by a flattened bag of fluid called a *disc*. These discs are each compressed by the 2100-N (newton) force and are subject to a pressure of $2100/A$ where A is their area. Typically A is about 5 cm² $= 5 \times 10^{-4}$ m² so the pressure is about 4×10^6 N/m² or about 40 times atmospheric pressure. This huge pressure (equivalent to a water depth of 400 m) is responsible for many of the problems of lower back pain arising from swollen discs that irritate the nerves.

In addition to the large compressive forces, the discs also must withstand substantial shear forces. These arise from the curvature of the spine necessary to give the erector spinae its 12° angle to work from. Without the curvature and consequent angle, it would be impossible for man to maintain balance, much less bend over.

C.   We now consider the mechanical structure of the human arm. The biceps muscle is attached to the forearm about 3 cm from the elbow joint (see Figure 1.5) resulting in a lever system with a mechanical advantage of about one-twelfth. This is a typical value for many joints and means that the force

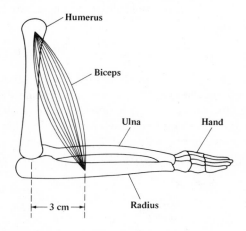

Figure 1.5 Diagram of the human arm. The drawing is only approximately to scale, and the only muscle shown is the biceps.

exerted by the hand from bicep contraction is 12 times smaller than the actual muscular force. In exchange for eleven-twelfths of the strength of our muscle we receive in return the ability to touch our shoulder with our wrist; if the biceps were connected directly from shoulder to wrist, we would be unable to feed ourselves. Because muscle can only contract by about 15% of its length, speed and flexibility can only be obtained from a system with mechanical advantage considerably less than unity.

Most muscle can produce a maximum force of about 50 N/cm² of cross section. Since the cross section of the average biceps is about 50 cm² (diameter of 8 cm $\cong$ 3 in.), the maximum force it can exert is about 2500 N, the weight of 250 kg, and the mechanical advantage of one-twelfth results in a force available at the hand of about 200 N (20 kg). This is the weight an average person can curl (lift with only the bicep as shown in Figure 1.5).

D. Figure 1.6 shows how the deltoid muscle acts when we extend an arm to the side. The center of mass of the arm is located about two-fifths of the

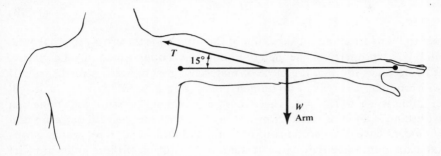

Figure 1.6 Diagram of the human torso with arm held extended. The drawing is only approximately to scale and the only muscle having any effect is the deltoid, which acts at a $\cong 15°$ angle to the humerus.

way from shoulder to hand, and if we consider the mass of the arm to act at that point, its weight W produces a clockwise torque. The arm is held still by two other forces. One of them comes from the tension T in the deltoid muscle, which acts at a 15° angle through a point about one-fifth of the way from shoulder to hand. The other force comes from the contact between the upper arm (humerus) and the shoulder bone (scapula). Since the forces at this ball-and-socket joint are not of immediate interest, we shall choose it as the origin so that they do not contribute to the torque equation. Then we find $2W/5 - (T \sin 15°)/5 = 0$ at equilibrium. From this we find $T = 7.7W$, well over 100 lb for a typical arm.

E. When a rectangular volume of material is subject to an applied force F, it will be compressed (or stretched) by an amount Δl. An elastic material is defined as one that stretches (or compresses) by an amount proportional to an applied force. If that force is doubled, Δl also doubles. Suppose we had two such pieces of material attached end to end and we applied a force. Each of them would be subject to the same force and would therefore stretch by the same amount. The total stretch would then be $2\Delta l$, twice as much. We say that the amount of stretch is proportional to the length.

If there were two pieces of material side by side, they would share the force F and therefore each of them would be subject to a force of $\frac{1}{2}F$. The resultant stretch would be $\Delta l/2$. We can see that the amount of stretch is inversely proportional to the cross-sectional area of the sample. The stretchiness or lack of it is described by a characteristic of the material called *Young's modulus* ($\equiv Y$).

The amount by which a sample will stretch is then

$$\Delta l = \frac{Fl}{AY} \tag{1.27a}$$

or

$$\frac{\Delta l}{l} = \frac{P}{Y} \tag{1.27b}$$

where l is its length in the direction of the force, A is the cross-sectional area, and Y is a property of the material known as Young's modulus. Of course, if Δl becomes too big, the piece of material can compress no further and it will crumble or break. For human bone, $Y \cong 2 \times 10^{10}$ N/m^2 and it will crumble when $\Delta l/l \cong \frac{1}{100}$. Therefore the maximum pressure that bone can withstand is about 2×10^8 N/m^2. Many load-bearing joints in our arms and shoulders have cross sections of only about 1 cm^2 = 10^{-4} m^2, which means the maximum force they can withstand is 2×10^4 N. If these joints are part of a leverage system with mechanical advantage of one-twelfth, the maximum forces that the limbs can exert are equivalent to the weight of 160 kg. It is clear that weight lifters and gymnasts who regularly submit their joints to much larger than ordinary forces must develop special tissue in their joints in order to avoid the danger of damaging them.

F. Normally when we jump or fall, we bend our knees to absorb the shock. What are the forces and pressures on the leg bones if we land with our knees locked straight? Our leg bones compress by an amount Δl that is related to the decelerating force F and Young's modulus of bone by $Fl = A Y \Delta l$ where A is the cross-section area and l is the bone length (see Eq. 1.27a). The acceleration F/m occurs over a distance Δl and is related to the height h of the jump or fall by (see Eq. 1.7)

$$gh = \frac{F\Delta l}{m}. \tag{1.28}$$

We therefore find

$$h = AY \frac{l}{mg}\left(\frac{\Delta l}{l}\right)^2 \tag{1.29}$$

Now the maximum value of $\Delta l/l$ that can be achieved without crushing is $\frac{1}{100}$ and the cross section of the bones in the leg is only about 4 cm^2. Since these bones are about $\frac{1}{2}$ m long, the maximim height from which a person can safely land this way is about $\frac{1}{2}$ m. Suprisingly enough, it's possible to break a leg by jumping off a chair.

If we land by flexing our legs instead of keeping them stiff, the forces are applied to the bones differently, but the total velocity change is the same. We can estimate the parameters used in Eq. 1.29 for the maximum safe height. The deceleration distance Δl is changed from $\frac{1}{200}$ m to about $\frac{1}{2}$ m (the s of

Part 1). However, force on the bones is increased by the reciprocal of the mechanical advantage of the muscle-bone structure. We must multiply the $\frac{1}{2}$ m by 100 for the increased Δl and divide by 12 for the mechanical advantage to find 4 m $\cong$ 12 ft as the height from which we might expect to be able to fall without breaking a bone. This agrees with common experience.

Part 4: Swimming

E. M. Purcell has examined the motion of a small swimming animal. Consider a bacterium swimming along under its own power by exerting a force to keep it moving against fluid friction. We ask what happens when it stops swimming and glides under the influence of its momentum. Since bacteria swim at slow speeds, the fluid friction force F that slows them is proportional to the velocity (see Appendix C) and is given by

$$F = -\beta v \qquad (1.30)$$

where β is the proportionality constant. The equation of motion is

$$m \frac{d^2x}{dt^2} = \beta \frac{dx}{dt} \qquad (1.31)$$

where m is the mass. We solve this equation subject to the initial conditions that the bacterium is at the origin at $t = 0$ when it stops swimming and that its velocity is v_0 at that point. We integrate the equation of motion and then try an exponential solution. The initial conditions require the solution

$$x = v_0\tau(1 - e^{-t/\tau}) \qquad (1.32)$$

where $\tau = m/\beta$ is the time constant for the slowing of the motion.

Observations of swimming protozoa under the microscope suggest that v_0 is about 10 body lengths per second or about 10^{-3} cm/sec. The parameter β can be evaluated for the special case of a spherical animal, which may not be a very bad approximation. Stokes' law for small particles moving in a fluid is

$$F = -6\pi\eta r \frac{dx}{dt} = -\beta v \qquad (1.33)$$

where η is the viscosity of the fluid (see Chapter 3) and r is the radius of the particle. The viscosity of water is about 0.01 Poise (the cgs unit of viscosity), so for a micron-size protozoan $\beta \cong 2 \times 10^{-5}$ gm/cm. If the density of the bacterium is approximately the same as that of water, its mass is about 4×10^{-12} g so that $\tau = m/\beta \cong 2 \times 10^{-7}$ sec. During an interval of τ sec the bacterium coasts a distance $0.63v_0\tau \cong 10^{-10}$ cm. These numbers are astounding because of their scale: the bacterium glides to a stop in a small fraction of a microsecond and traverses about $\frac{1}{100}$ of an atomic diameter in the process.

APPENDIX A
CONSTANT MUSCLE FORCE

We now consider the implications of the approximation that muscles exert a constant force during acceleration. Various experiments have shown that the force a muscle exerts is a maximum at its relaxed length and is less than this at both shorter and longer lengths. The curve of maximum muscular force versus length x is approximately parabolic (see Figure A.1) and can be described approximately by the equation

$$F = F_0 - \Delta F \left(\frac{x - x_0}{x_m} \right)^2. \tag{A.1}$$

Here F_0 is the maximum force the muscle can exert at its relaxed length x_0, and $F - \Delta F$ is the maximum force the muscle can exert at maximum and minimum length $x_0 \pm x_m$.

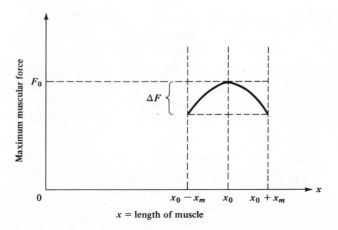

Figure A.1 Plot of maximum muscular force versus muscle length. This curve is characteristic of single muscle fibers (the myofibrils) as well as whole muscles. Note that the muscle can only change its length about 20%.

We wish to calculate the amount of work that a muscle with such a force curve can do and to compare that with the amount of work we obtain by assuming a constant force whose magnitude is given in a prescribed way. We choose a constant force that is equal to the average of the minimum and maximum values. (Of course we can always choose a value of the constant force so that the answers for the work will come out the same: we must choose one that does not assume that we already know the answer.) This average force is $F_{av} = \frac{1}{2}(F_0 + F_0 - \Delta F)$ and we find that the maximum work done by it in moving from $-x_m$ to x_m is $W_0(1 - \Delta F/2F_0)$ where $W_0 = 2x_m F_0$.

The actual work done by the force is given by the integral

$$W = \int_{x_0-x_m}^{x_0+x_m} F \cdot dx = \int_{x_0-x_m}^{x_0+x_m} \left[F_0 - \Delta F\left(\frac{x-x_0}{x_m}\right)^2 \right] dx$$

$$= 2x_m F_0 - \frac{2\Delta F}{3} \frac{x_m^3}{x_m^2}. \qquad (A.2)$$

Then we find

$$W = W_0\left(1 - \frac{\Delta F}{3F_0}\right). \qquad (A.3)$$

The difference between the approximation $W_0[1 - (\Delta F/2F_0)]$ and Eq. A.3 is $W_0 \Delta F/6F_0$. Note that, for typical cases, $\Delta F \cong F_0/4$ so that the approximation causes errors that are less than 5%.

APPENDIX B
BALLISTIC DISTANCE

We wish to calculate the maximum distance d a ballistic object can travel. In ballistic motion (neglecting air friction) the vertical component of velocity is $v_{0y} - gt$ where v_{0y} is the initial y velocity. When the object returns to the ground, its vertical velocity is just equal but opposite to v_{0y} so the time of flight t_f is found from

$$v_y = v_{0y} - gt_f = -v_{0y} \tag{B.1}$$

or $t_f = 2v_{0y}/g$. The distance traveled is given by

$$d = v_x t_f = \frac{2v_x v_{0y}}{g} \tag{B.2}$$

where v_x is the (constant) horizontal component of velocity. For an object projected with velocity v at an angle θ above the horizontal

$$v_x = v \cos \theta \quad \text{and} \quad v_{0y} = v \sin \theta. \tag{B.3}$$

Then Eq. B.2 becomes

$$d = \frac{2v^2}{g} \cos \theta \sin \theta = \frac{v^2}{g} \sin 2\theta. \tag{B.4}$$

The maximum value for d occurs when $\theta = 45°$ and is

$$d = \frac{v^2}{g}. \tag{B.5}$$

APPENDIX C
FLUID FRICTION

The problem of the force exerted on an object moving in a fluid is extremely complicated. The general case has not yet been solved, and many special cases have only partial or approximate solutions. The solutions depend on the speed of the object, the shape of the object, whether or not the fluid is compressible, and a number of other variables. Furthermore, the dependence on some of these variables is often not analytic: A slight change in some parameter may result in a drastic change in the nature of the fluid friction. The development of turbulence or boundary layer separation are examples of discontinuous behavior.

In order to make any quantitative description of this problem, it is necessary to use some description of fluids. It will be assumed here that the reader is familiar with the contents of Chapter 3. We shall consider the motion of a spherical object in a stationary fluid subject to viscous and inertial forces. The viscous forces arise from the friction of the fluid as it rubs along the sides of the object, and the inertial forces arise because the moving object has to push some of the fluid out of the way.

We make the simplifying assumption that the fluid is essentially undisturbed at a distance of about one radius from the moving object and that the fluid in contact with the object is moving with it. The viscous force is given by $F = \eta A \, (dv/dx)$ where η is the viscosity of the fluid, A is the area of the object along which the fluid drag force takes place (the normal to this surface is perpendicular to the velocity), and dv/dx is the velocity gradient in the fluid perpendicular to the object's velocity. This formula is discussed in Chapter 3.

In order to provide some simple way to apply this formula to our problem, we approximate (dv/dx) by v/r since Δv from object to stationary fluid is simply v and we assume the distance to the stationary fluid is approximately r. We also approximate the area by $4\pi r^2$ even though the surface of a sphere is round and therefore has no finite area whose normal is perpendicular to the motion. These approximations lead to the expression $F = 4\pi\eta rv$, which turns out to be correct except for a proportionality factor. The fluid friction on a spherical object is correctly given in a limited range of speed by Stokes' law as

$$F = 6\pi\eta rv. \tag{C.1}$$

It is necessary to compare this force with the inertial force. From Bernoulli's law (see Chapter 3) we find that the pressure difference on a streamline that skims the front surface of the moving sphere is $\frac{1}{2}\rho v^2$ between the fluid near the sphere and that at some distance away. This is because the fluid at the surface is moving at velocity v (with the object) but the fluid far away is stationary ($v = 0$). The force produced by this pressure difference is found by multiplying by the area, but this time the area is the projection whose normal is parallel to the velocity. In the special case of the sphere, this is πr^2 and we find that the force is

$$F = \frac{\pi}{2} \rho(rv)^2. \tag{C.2}$$

Note that the r's that appear in Eq. C.1 and C.2 are only the same for spherical objects. This expression, like Eq. C.1, is only valid in limited regions.

We see that the two expressions depend on the velocity differently, and we shall evaluate the coefficients in order to compare them. The dimensionless ratio of these two forces is $\rho v r/\eta$ (omitting constants), which is the Reynolds' number N_R discussed in Chapter 3. Consider the case of a spherical object falling through air of density 1.3×10^{-3} g/cc and viscosity 1.8×10^{-4} Poise (see Chapter 3). Equation C.1 becomes $F = 3.4 \times 10^{-3} \, (rv)$ and Eq. C.2 becomes $F = 2 \times 10^{-3}(rv)^2$ from which it is clear that for values of rv large compared with 1 cm²/sec the inertial force dominates and for values of rv small compared with 1 cm²/sec the viscous force dominates.

We shall evaluate three separate cases. In Part 2E on p. 7 we discussed the terminal velocity of falling objects. We set the fluid drag force equal to the weight and use Eq. C.2 to find the terminal velocity. For a human body of radius $r = 30$ cm and mass 75 kg we find $v_t = 64$ m/sec, which is not very different from that experienced by parachutists. In this case rv is about 200,000 cm²/sec, which implies that we have correctly chosen the inertial force for the fluid resistance. For a raindrop of radius 0.1 mm, the inertial formula gives $v_t = 140$ cm/sec and the viscous formula gives $v_t = 120$ cm/sec. For this raindrop, rv is of order unity. For a 50-micron diameter dust particle of density 1, we find $v_t = 7.7$ cm/sec. In this case $rv = 0.02$, which implies that we have correctly chosen the viscous force for the fluid resistance.

In order to make similar comparisons for motion through other fluids, it is necessary to re-evaluate the constants in Eqs. C.1 and C.2. For example, in water the viscous and inertial forces become equal for $rv = 0.13$ cm²/sec.

It must be stressed that this appendix has only served to provide a description of fluid friction under very special circumstances. The descriptions given here sometimes break down for cases where there is turbulence, appreciable compressibility, boundary layer separation, entrainment of particles, and where the moving particles become comparable in size to the molecules of the fluid. The size and shape of an object can either suppress or enhance the onset of turbulent flow, which is important in both viscous and inertial cases.

Questions and Problems for Chapter 1

1. What is the correct expression relating the potential energy and the height reached by a ballistic object for the case when h is not small compared with the radius of the earth? Neglect air friction.

2. It is quite possible for an animal to leave the ground without raising his center of mass at all. Draw a diagram to show how this occurs. Describe the motion of a person running a hurdle race.

3. When a well-trained vaulter jumps over a bar, his center of mass usually passes under the bar. (Obviously one can clear a much higher bar this way.) How is this possible? Draw a diagram to explain your answer.

4. If we apply Eq. 1.12 to a golf ball, it should be possible to throw it more than half of a mile. This is absurd. We have left out air friction in this discussion, and it would certainly tend to reduce the half-mile estimate. Even if air friction were included, however, this analysis would yield a ridiculously large result for a golf ball. Why? What has been left out? How does a golf club work?

5. Equation 1.12 gives good results for the 16-lb shotput but nowhere near good results for the 16-lb hammer throw. Why is that? (The 16-lb hammer is a shotput on the end of a 3-ft chain.) What can we say about javelin and discus throwing (running throw and spinning throw)?

6. How would Eq. 1.12 have to be changed in order to account for the fact that the launch of the projectile is above ground level but the end of the flight is at ground level?

7. A pole vaulter raises his center of gravity by at least 4 m, and he has no source of energy other than his own muscles. Why does the scaling argument not apply and limit him to about 1 m? How does the maximum height of a pole vault scale with size?

8. Assuming a pole vaulter can run at 10 m/sec, how high a bar should he be able to clear? How does this result compare with world records? What is the effect of elastic poles?

9. Try to estimate the error involved in calculating terminal velocity v_t by ignoring air friction for velocities less than v_t.

10. What size must an animal be in order to jump high enough to land at terminal velocity? What can you say about launch speed in this case?

11. Using arguments similar to those used for scaling of pulse rates, calculate respiration rate scaling.

12. What size must an animal be in order to require its own weight in food daily? Can you guess why there are very few small animals living in cold climates?

13. McMahon's argument required that surface area scale as l^3, not l^2. Show that heart rate still scales as l^{-1}.

14. Why do small animals have such a high pulse rate? Since they have less blood to pump around than a large animal, it seems they could get by with a lower pulse rate.

15. A shark cannot pump water over its gills and so must keep moving in order to have a supply of oxygen. Assume the resistive force of the water when the shark is moving is proportional to the cross-section area and the velocity. Make some reasonable assumption about the way the shark's oxygen supply and need scales with size, and determine how swimming speed should scale with size.

16. When a particular steel ball of radius r is dropped into a column of water, it reaches a terminal velocity of 10 m/sec. What is the terminal velocity for steel balls of radius $2r$ and of radius $r/2$? What is terminal velocity of a copper ball of radius r? What about a cork ball of radius r?

17. Suppose you are being chased by a large (angry) bear who can run as fast as you can on the level ground. Use a scaling argument to show that you should run uphill to get away.

18. What would happen to the world as we know it if all objects were scaled larger 100X in linear dimension. Discuss the effects on living systems, both plants and animals. Give examples.

19. Repeat the calculations for F_e and F_p (Eq. 1.25) under the conditions of lifting a weight with the arms. This is the position often assumed when a mother lifts a child who might have a mass of 15 kg (about 30 lb). Notice how this modest force is multiplied by the structure of the skeleton. It should be obvious why we should lift weights by bending our knees and keeping our backs straight rather than vice versa.

20. Make a reasonable assumption about the area of a spinal disc (between vertebrae) and calculate the pressure on it under conditions similar to those in Exercise 19.

21. Assume that the vertebrae are stacked and tilted as shown in the diagram in order to achieve the curvature. Calculate the shear force (perpendicular to compression) on the discs assuming the trunk, head, and arms constitute $0.6W$.

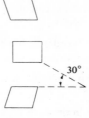

22. Estimate the moment of inertia of the forearm pivoted about the elbow and calculate the maximum velocity the hand can reach if it is moved only by the forearm. How high can you throw a light object this way?

23. Derive Eq. 1.28 using the work-energy theorem.

24. In the accompanying diagram there is a drawing of the knee. You are asked to do knee calculations similar to those done in Part 3C for the elbow. Make reasonable assumptions about dimensions; using the conditions for equilibrium, find an equation for the external force that is produced (F_e) when there is a given contact force F_c between the bones. Use $d = 5$ cm. Use the crushing strength of bone (10^8 N/m²) to determine the maximum force that could be exerted by the bent knee. (Use area $A = 3$ cm².) Now take $a = gh/s$ and determine the maximum height from which a person might fall safely without breaking a bone. Remember there are two legs to break the fall.

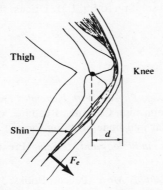

25. Calculate the force exerted by the scapula on the humerus (see Figure 1.6) when the arm is held straight out. Estimate the cross-section area of the joint and calculate the pressure.

26. The deltoid muscle is attached to the middle of the humerus and acts at an angle of about 15° when the arm is outstretched horizontally as shown in Figure 1.6. Calculate the tension in the deltoid muscle for the case where the arm supports a gallon of wine (5 kg, including bottle).

27. When standing on the toes of one foot as in climbing stairs, the entire body weight is supported by the force in the Achilles' tendon as shown in the diagram.

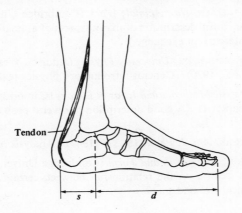

Draw a vector diagram showing the forces. By feeling around your own foot, make reasonable estimates of s and d, and calculate the force on the tendon. Estimate the diameter of the tendon and calculate the stress (force/area) on it. Compare this with atmospheric pressure $= 10^5$ N/m^2.

28. It is known that the human bicep can exert a force in excess of 500 lb. Why can't we lift this weight?

29. In Figure 1.3 it is clear that humans are anomalously long lived in the animal world. Why do you suppose that is so?

30. Show that Eq. 1.32 is a solution to Eq. 1.31. Discuss the choice of the constants.

Bibliography

1. G. BENEDEK and F. VILLARS, *Physics With Examples From Medicine and Biology*, Vol. I (Addison-Wesley Publishing Co., Inc., Reading, Mass., 1974). (Contains an excellent discussion of shear and bone-breaking forces.)

2. PAUL DAVIDOVITS, *Physics in Biology and Medicine* (Prentice-Hall, Inc., Englewood Cliffs, N.J., 1975).

3. SIR JAMES GRAY, *How Animals Move* (Pelican Books, Harmondsworth, England 1959). (A delightful description of a variety of animal movements including swimming, slithering, walking, etc.)

4. RUSSELL HOBBIE, Intermediate Physics for Medicine and Biology, John Wiley & Sons, N.Y. 1978. (A broad text which discusses many related topics.)

5. T. MCMAHON, *Science*, 1201 (March 1973). (A good discussion of several scaling problems.)

6. E. PURCELL, *Am. J. Phys.*, **45**, 3 (1978). (Life at low Reynolds numbers.)

7. R. RESNICK and D. HALLIDAY, *Physics* (John Wiley and Sons, Inc., New York, 1968). (General introductory physics text.)

8. KNUT SCHMIDT-NIELSEN, *How Animals Work* (Cambridge University Press, New York, 1972). (A delightful description of many aspects of animal physiology. Special emphasis on countercurrent exchange.)

9. F. SEARS and M. ZEMANSKY, *University Physics* (Addison-Wesley Publishing Co., Inc., Reading, Mass., 1970). (General introductory physics text.)

10. J. MAYNARD SMITH, *Mathematical Ideas in Biology* (Cambridge University Press, London, 1968, Chapter 1). (A good description of several problems related to classical biophysics.)

11. S. C. STRAIT, et al., *Am. J. Phys.*, **15**, 375 (1947). (Analysis of skeletal forces.)

12. D. W. THOMPSON, *On Growth and Form* (Cambridge University Press, London, 1917). (A famous treatise on the relationship between certain physiological phenomena and physical size and shape.)

13. M. WILLIAMS and H. LISSNER, *Biomechanics of Human Motion* (W. B. Saunders, Co., Philadelphia, Pa., 1962).

14. DAVID WILLOUGHBY, *Natural History Magazine*, 57 (December 1969). (Contains data on animal longevity.)

15. ———, *Natural History Magazine*, 69 (March 1974). (Contains data on animal running speeds.)

2

HEAT
AND
ENERGY

Part 1 : Heat and Calorimetry—A Review

The science of mechanics, with the attendant notion of energy conservation, was reasonably well developed by the eighteenth century in the work of LaGrange. It wasn't until the middle of the nineteenth century, however, that heat was recognized as a form of energy. Until that time, heat had been considered as some sort of fluid or other ethereal substance. The work of Joule and Kelvin established the relationship between heat and mechanical energy.

Heat plays a very important role in the energetics of the body. It is necessary to maintain body temperature, to promote metabolism, and to provide the proper thermal environment for many chemical processes. Heat produced by muscular action as well as by other metabolic processes must be properly controlled so that the temperature is kept stable.

In order to introduce certain notions about heat we shall study some thermal properties of gases. We begin with Newton's laws embodied in the equation

$$\vec{F} = \frac{d\vec{p}}{dt} \tag{2.1}$$

which says the force $\vec{F}$ is the time rate of change of momentum $\vec{p}$. We assume that a gas can be treated as a swarm of very small, hard spheres and that the velocities of these molecules are equal in magnitude but uniformly distributed in all directions. We consider the momentum transfer to a wall in the yz plane from those molecules with a positive x-component of velocity and calculate the pressure as follows. In a cylindrical volume whose axis is perpendicular to the wall (see Figure 2.1), one-half of the molecules within the distance $v_x t$ will strike the wall within time t (the other half of the molecules are going away from the wall). The total number of molecules striking an

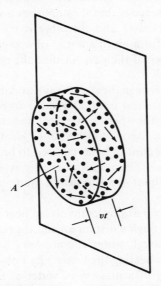

Figure 2.1 A selected volume of gas molecules near a wall. The cylindrical volume has base area A and height $v_x t$.

area A of the wall in time t will be $\frac{1}{2}(n/V)(Av_x t)$ where n/V is the number per unit volume. The force F they exert is

$$F = \frac{dp}{dt} = \frac{1}{2}\frac{n}{V}(Av_x)(2mv_x) = \frac{n}{V}Amv_x^2 \qquad (2.2)$$

since the momentum transfer from the molecules to the wall is $2mv_x$ per molecule $(v_x \longrightarrow -v_x)$. The pressure P is the force per unit area and we find

$$PV = nmv_x^2. \qquad (2.3)$$

Since we expect

$$v_x^2 = v_y^2 = v_z^2 \qquad (2.4a)$$

and

$$v_x^2 + v_y^2 + v_z^2 = v^2 \qquad (2.4b)$$

we find

$$PV = \tfrac{2}{3}n(\text{K.E.}) \qquad (2.5)$$

where K.E. $= \frac{1}{2}mv^2$ is the average kinetic energy of each molecule.

We now consider the relation between Eq. 2.5 and Boyle's law and Charles' law for ideal gases. These laws give us the ideal gas equation

$$PV = nkT \qquad (2.6)$$

where $k \cong 1.4 \times 10^{-23}$ J (joule)/°K is the Boltzman constant and T is the absolute (Kelvin) temperature. Comparison of Eq. 2.5 and Eq. 2.6 gives a relationship between temperature and energy that is very important. The major reason for its importance is that it establishes a value for the mechanical equivalent of heat for gases and therefore it tells us how much of a temperature rise to expect for a given amount of energy input. The quantity dE/dT, where E is the energy of the gas, is called the *specific heat C* and is a constant $\frac{3}{2}k$ for an ideal gas.

Similar relationships between heat (energy) and temperature exist for other materials, but they are not easily derivable from first principles. We therefore will simply state the specific heats of various materials and use them in equations involving energy conservation. Heat energy is commonly measured in terms of the amount of heat required to heat 1 gram of water by 1 degree (Celsius *or* Kelvin). The required energy is 4.18 J and is called a *calorie*. Therefore it takes about 418 J to heat a gram of water from freezing to boiling (0° to 100°C).

We now proceed to do two examples involving heat and energy conservation. Suppose water falls through a distance h and the resultant kinetic energy is converted to heat. (Although waterfalls are very noisy, the fact is that most of the energy goes into heating the water.) By how much does the water warm up? The total energy of a mass m of water at the top of the falls is $mgh + mCT_1$* where C is the specific heat of water, one calorie/per gram. The kinetic energy at the bottom of the falls is converted to heat so that total energy is all thermal and is simply mCT_2. Then the temperature rise

$$T_2 - T_1 = \Delta T = \frac{gh}{C}. \qquad (2.7)$$

The specific heat of water is 1 cal/g °K = 4.18 J/g °K = 4180 J/kg-°K. Then we have

$$\Delta T = \frac{g \ (\text{m/sec}^2) \ h \ (\text{m})}{C \ (\text{J/kg-°K})} = \frac{9.8h}{4180} \ °\text{K} \cong \frac{h}{425} \ °\text{K} \qquad (2.8)$$

which means a 425-m falls is high enough to heat the water by 1°. The great upper Yosemite falls is the only steady one in the United States that is high enough to raise the water temperature by 1°.

If 20 g of cream at 0°C is added to a 200-g cup of coffee at 80°C, what is the final temperature of the mixture? We assume that the specific heat of

---

*This is the total energy only if C is constant for all temperatures between 0 and T, which is usually not the case. We shall always deal with a restricted temperature range in which C is constant.

both coffee and cream are equal to that of water, and write an energy equation. We find

$$\text{initial energy} - \text{final energy} = 0 \tag{2.9a}$$

$$20(273) + 200(353) - 220(T) = 0 \tag{2.9b}$$

from which $T = 346°K$ or $73°C$.

Part 2 : Heat Transfer

The three mechanisms for heat transport are conduction, radiation, and convection. We shall begin with a discussion of conductive heat transfer across a boundary. Once again we choose to study a gas since it is easy to calculate things from first principles. In Appendix D we show that, in a head-on collision of equal masses where there is no loss of mechanical energy, the relative velocity of the two bodies only changes sign in the collision. That is, if two molecules are traveling with velocities v_1 and v_2, then

$$(v_1 - v_2) = (v_2' - v_1') \tag{2.10}$$

where primes represent velocities after a collision. Since we consider collisions between molecules of equal mass, conservation of momentum gives

$$v_1 + v_2 = v_1' + v_2'. \tag{2.11}$$

We combine Eq. 2.10 and 2.11 by first adding the equations and then subtracting them and see that the molecules simply exchange velocities. We find

$$v_1 = v_2' \tag{2.12a}$$

and

$$v_2 = v_1'. \tag{2.12b}$$

We notice that if v_1 is larger than v_2, then it must be that v_2' is larger than v_2. That is, if molecule 1 comes from a region of gas that is hotter (i.e., moving faster) than molecule 2, molecule 2 is heated (i.e., made to go faster) in a collision. Hot molecules transfer heat to colder ones. Heat energy flows from hot to cold!

We also notice that molecule 2 is sent back into its original colder region somewhat warmed up (see Fig. 2.2a). It carries heat back there which it transfers to colder molecules and heats them up. In the course of this action it may be returned to the boundary layer of region 1 and be heated up again [Figure 2.2(b)]. Furthermore, the molecule in region 2 (cold) that has been warmed up is sent deeper into region 2 to carry heat still further. We see how heat is diffused from a hot region to a colder region by means of collisions.

This process is not restricted to gases. Heat flows in solids and liquids by virtually the same mechanism. We can describe the heat flow quantitatively

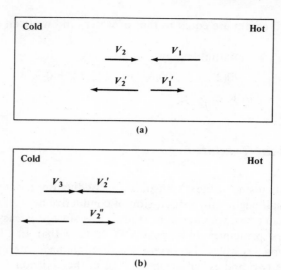

(a)

(b)

Figure 2.2 Schematic illustration of head-on collisions between molecules. In (a) a fast molecule collides with a slower one transferring energy to it. In (b) the molecule that was struck has traveled into a cooler region of the gas and deposits some of its energy there.

as follows. If opposite ends of a piece of material are kept at different temperatures, the amount of heat energy that flows from the hot end to the cool end is proportional to the temperature difference ΔT, to the cross-section area A of the piece of material, and to the time Δt for which the heat is allowed to flow. The amount of heat ΔQ is also inversely proportional to the distance d over which this heat must flow. The proportionality constant K is called the *thermal conductivity* and we write

$$\frac{\Delta Q}{\Delta t} = \frac{KA\Delta T}{d}. \qquad (2.13)$$

We can use this equation to calculate the heat lost by the human body to the ambient air. We make the simplifying assumption that there is a layer of subcutaneous fat that serves to insulate the body against heat loss and that this fat is of uniform thickness everywhere. Furthermore, we assume that the temperature on the inside of this layer is 37°C everywhere and that the skin temperature on the outside of this layer is 30°C. The amount of heat that flows through this fat layer is given by Eq. 2.13 where ΔT is 7°C; d, the thickness of the fat, is 1 cm; and $K_{fat} = 5 \times 10^{-4}$ cal/cm-sec-°C. We consider the body as a cylinder of 1 m circumference and 1.5 m long and estimate that the surface area of the body is about 1.5 m² = 15,000 cm² and then we find $\Delta Q/\Delta t = 53$ cal/sec (a "food calorie" = 1 kcal). In a 24-hour period we lose 4.6×10^6 cal or 4600 kcal! This is an astoundingly large number in view of the fact that most adults only metabolize approximately 3500 kcal/day.

The resolution of the difficulty lies in the fact that we wear clothing over most of our bodies for most of the day. This clothing traps a layer of air approximately 1 cm thick all around our bodies.* Since the thermal conductivity of air is 6×10^{-5} in the same cgs units, the clothing reduces the heat loss by a factor of about 10. Of course, a thicker layer of clothing or hair will insulate us against a colder environment.

Radiative heat transfer from the human body can best be described by the Stefan-Boltzman law for thermal radiation from a nonreflecting object:

$$P = \sigma A T^4, \qquad \sigma \cong 5.67 \times 10^{-8} \text{ W/m}^2\text{-}°\text{K}^4 \qquad (2.14)$$

where P is the total power radiated by a body of area A and surface temperature T (°K). The spectral distribution of this thermal radiation (see Chapter 7 for a discussion of spectra) was calculated many different ways, but only Planck's famous quantum formula agreed with the measured spectrum. The energy density (energy per unit volume) U at wavelength λ per unit wavelength range is

$$U = \frac{8\pi ch}{\lambda^5} \frac{1}{(e^x - 1)}, \qquad x = \frac{hc}{\lambda kT} \qquad (2.15)$$

where $h \cong 6.6 \times 10^{-34}$ J-sec is Planck's constant. (The derivations of Eq. 2.14 and 2.15 are based on thermodynamic considerations beyond the scope of this book.) This spectrum has a rather broad peak centered at wavelength λ_p given by Wien's law:

$$\lambda_p \cong \frac{hc}{5kT} \qquad (2.16)$$

which is about $10\,\mu$ (infrared radiation) for body temperature. Note that small variations in T produce small variations in λ_p. Infrared thermography is a widespread diagnostic method that utilizes these small differences.

If we take $T_b = 310°$K (37°C—body temperature) and use Eq. 2.14 to calculate the power radiated by a human, the result is about 10 times an adult's daily consumption of energy. In order to resolve this discrepancy, we must remember that we live in an environment at $T_r = 293°$K (20°C—room temperature) and that environment radiates energy to us. Furthermore, the temperature at the skin, which is the surface that does the radiating, averages closer to $T_s = 305°$K. We then find that the net power radiated is

$$P_{\text{radiated}} = \sigma A(T_s^4 - T_r^4) \cong 4\sigma A T_r^3(T_s - T_r) \cong 1800 \text{ kcal/day.} \qquad (2.17)$$

---

*Clothing is mostly air. To see this easily, fill a graduate or beaker with water up to some some calibration point and then submerse an article of clothing such as a T-shirt in it. After all the air has escaped and the clothing is thoroughly soaked, observe how high the water has risen. The small displacement of the water level indicates that most of the volume of the clothing is air.

The effect of nonuniform skin temperature and insulation from clothing reduces the radiative heat loss to some extent, but not by so much as it reduces conductive heat loss. It appears that a significant part of the energy consumed by a human is used to maintain body temperature against radiative heat loss.

Convective heat transfer is the most difficult to discuss quantitatively because there are so many unknown variables. A person unprotected by clothing will subject his skin to the local environmental temperature resulting in rapid heat loss. This lost heat may be carried away by air or water flow so fast that the skin is lowered to the environmental temperature. Although this reduces radiative heat loss, it increases conductive heat loss unless the conductivity from the warm body core to the peripheral areas is reduced. We shall see that the body has a mechanism to do this.

Now that we have discussed various processes of heat transfer, it is appropriate to discuss a phenomenon called *countercurrent heat exchange*. This process occurs when two fluids at different temperatures are flowing in opposite directions in separate tubes that are in thermal contact with one another. In Figure 2.3 we see that if the warm fluid enters the left-hand tube at a temperature of 40°C, it will lose some of its heat by conduction to the cooler fluid in the right-hand tube and further downstream it will be cooled

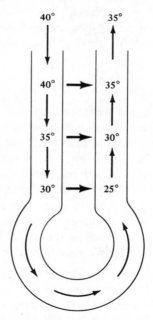

Figure 2.3 Illustration of a thermal countercurrent exchange loop. Warm fluid enters at the upper left and leaves at the upper right only slightly cooler than its entering temperature, even though it was much cooler near the bottom of the loop. Horizontal arrows indicate heat flow.

to some lower temperature, say 35°C. At the same time, the cooler fluid will be warmed up by absorption of the transferred heat. In the example shown, it will heat up from 30° to 35°C. The same process occurs everywhere along the region where the tubes are in contact: In the lower region the warm fluid cooled to 35°C is cooled still further to 30°C while the cooled fluid entering at 25°C is warmed to 30°C. For the particular case illustrated here, the two counterflowing fluids are actually the same one, flowing in a closed loop. We see that some heat is lost right at the bottom of the loop corresponding to a temperature drop of 5°C, from 30°C to 25°C. The phenomenon of counter-current heat exchange allows a region (at the bottom of the loop) served by a fluid to be cool without resulting in a substantial heat loss to the fluid. In this example, the fluid returns only 5°C cooler than the temperature at which it left (35° versus 40°C) but serves a region that is 15°C cooler. Two-thirds of the heat that might have been lost is transferred horizontally as shown by the thick arrows.

In Appendix E it is shown that the thermal gradient and rate of heat transfer are both constant along uniform tubes or vessels. Of course, this is what might be intuitively expected on the basis of conservation of energy, but the derivation can be modified for nonuniform flow or thermal conductivity. The results support our notion that countercurrent heat exchange is enhanced by proximity, slow flow, and good thermal contact.

Structures that permit countercurrent exchange are very common. The flippers of whales, which are warm-blooded, are served by arteries that are surrounded by veins. If a whale is in cold water, the large surface area of the flippers are ideal for promoting substantial heat loss from the warm core of the animal. If blood were returned to the core from the flippers as cold as the surrounding water, the animal's metabolism would have to provide the heat to warm it up. With the countercurrent exchange process functioning, the warm arterial blood from the core warms the returning blood in the veins and is cooled substantially in the process. When the cooled blood serves the tissue in contact with the cold water, it cannot lose as much heat as it would have if it were at core temperature. The burden on the metabolism is considerably reduced.

The vasculature of the human forearm is similarly constructed to allow for countercurrent heat exchange. The process is a little bit more complicated than the simple case of the whale flipper. In the forearm there are two sets of veins (see Figure 2.4). One of these sets of veins lies close to the arteries and exchanges heat with them in exactly the way we have described. The other set lies close to the surface where it can exchange heat easily with the surrounding air. The returning blood is steered to the inner or outer veins by vasodilating and vasoconstricting muscles that are controlled by temperature sensors in the hand. If the hand is in a cold environment, blood is returned through the inner veins and heat loss is reduced by countercurrent exchange. We have all experienced the "cold hands" phenomenon that results from this process. If

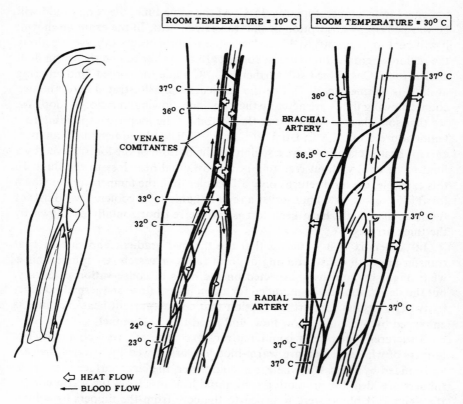

Figure 2.4 Illustration of countercurrent exchange mechanisms for heat in the vasculature of the human forearm. The open arrows indicate the heat flow and the single-line arrows indicate blood flow. (From Selkurt, *Physiology*, 3rd ed. © 1971 Little, Brown, and Co. Reprinted by permission.)

the environment is warm, however, the blood is caused to flow near the surface and heat is lost to the environment resulting in a cooling of the body. This very beautiful and complex system serves the dual function of preserving heat when it's needed and of cooling the body otherwise.

The process of countercurrent exchange can be used to cool part of the body deliberately. The testes of mammals are most fertile when they are a few degrees below body temperature, and it is for this reason that nature has exposed them to the cooling of the surrounding air. The fertility of many mammals drops substantially when the testes temperature is raised even a few degrees. This anatomical device would fail, however, if the testes were continuously infused by warm blood from the core. The countercurrent exchange process allows them to be kept cool and fertile with little loss of heat from the body. Countercurrent exchange also permits animals whose body temperatures may rise substantially from violent exercise to maintain their brains at tolerable temperatures.

Many animals have a well-developed structure called *rete mirabile* (wonderful net) to promote countercurrent exchange. These retes, often located near the junction between a limb and the trunk of an animal, consist of many bundles of small arteries and veins close to one another. The vessels are considerably larger than capillaries and therefore do not permit substantial oxygen exchange, but they do trap a great deal of heat in the animal's body. Some arctic animals have retes so effective that their limb temperatures are as low as 10°C.

The principle of countercurrent exchange is not limited to heat conservation. We can understand its utility in a variety of situations by recognizing that, in the heat exchange systems we have considered so far, a small thermal gradient across the vein-artery boundary produces a large temperature difference across the ends of the structure.

In a system characterized by active transport, there is usually some maximum gradient that the transport mechanism can sustain. The countercurrent process can function as a multiplier under conditions of active transport and enhance the gradient considerably. In Figure 2.5 there is active transport by

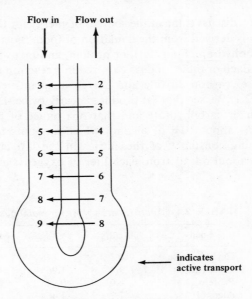

Figure 2.5 Schematic diagram of concentration enhancement in a countercurrent exchange loop using active transport. The maximum gradient the active transport can achieve is 1, but the concentration difference between loop ends is 6.

the separating membrane in the direction indicated by the arrows. We see that the membrane performing active transport along the length of the countercurrent exchange structure is supporting a concentration difference of 1 unit everywhere, but the concentration difference from one end of the structure to the other is 6 units. This "multiplier" effect is found in a variety of animal functions including the Henle's loop of the nephrons in mammalian kidneys (for concentration of urine) and the swim bladders of deep-sea fishes (to obtain gas at high pressures).

Whether the transport across the separator is active or passive, the efficiency of the transport is very high. Nature has taken advantage of this beautiful system in a large number of unrelated structures. For example, in bird lungs the air flow, which is unidirectional rather than oscillatory, is opposite to blood flow in order to ensure the most effective use of the inhaled oxygen. Another example is found in the respiratory and urinary systems of desert animals that are designed to conserve water.

Part 3:   Energy from Metabolism

We now discuss some of the sources of energy for the human body. Muscular energy is derived from the oxidation of food, most of which can be classified as carbohydrate, lipid (fat), or protein, and many foods are a mixture of all three. Each of these foods is oxidized in a reaction that consumes oxygen and produces carbon dioxide and energy in quantities given approximately by Table 2.1. We see that fat produces nearly twice the energy per gram as does protein or carbohydrate and that one ounce of 86 proof liquor ($\cong 30$ cc) contains about 10 g of alcohol which amounts to 71 kcal. Also, note the surprising constancy of the entries in the last column, which shows that we get about 5 kcal from each liter of oxygen we consume no matter what we eat.

TABLE 2.1 *Approximate values for metabolism of most foods.*

Food	Kcal/g	$\ell\,O_2$/g	$\ell\,CO_2$/g	Kcal/$\ell\,O_2$
Carbohydrate	4.1	0.81	0.81	5.0
Lipid (fat)	9.3	1.96	1.39	4.7
Protein	4.0	0.94	0.75	4.5
Alcohol	7.1	1.46	0.97	4.9

Note that these numbers can vary substantially over different choices of foods in the same category.

The oxidation reactions for typical foods are shown below:

Sugar

$$\underbrace{C_6H_{12}O_6}_{180\,g} + \underbrace{6\,O_2}_{134.4\ell} \longrightarrow \underbrace{6\,CO_2}_{134.4\ell} + \underbrace{6\,H_2O}_{108\,cc} + 686\ \text{kcal/mole}$$

$$\tfrac{686}{180} = 3.8\ \text{kcal/g.} \qquad (2.18)$$

Fat

$$\underbrace{(CH_2)_{52}(COO)_3}_{860\,g} + \underbrace{78\,O_2}_{1747\ell} \longrightarrow \underbrace{55\,CO_2}_{1232\ell} + \underbrace{52\,H_2O}_{936\,cc} + 8160\ \text{kcal/mole}$$

$$\tfrac{8160}{860} = 9.5\ \text{kcal/g.} \qquad (2.19)$$

Alcohol

$$\underbrace{C_2H_5OH}_{46\,g} + \underbrace{3\,O_2}_{67.2\ell} \longrightarrow \underbrace{2\,CO_2}_{44.8\ell} + \underbrace{3\,H_2O}_{55\,cc} + 327\ \text{kcal/mole}$$

$$\tfrac{327}{46} = 7.1\ \text{kcal/g.} \qquad (2.20)$$

The normal adult is active enough to metabolize about 3500 kcal/day. The last column of Table 2.1 shows that in order to do this he needs about 700 ℓ of oxygen, which is contained in about 3500 ℓ of air (air is 20% O_2). This means that an adult needs about 150 ℓ of air per hour; however, our half-liter breaths occur about 12 per minute so we breathe 360 ℓ per hour. It is clear that we must exhale more than 50% of the oxygen we inhale.

The constancy of the right-hand column of Table 2.1 allows certain measurements to be made rather easily. In order to determine basal metabolism rate (rate of energy production while completely at rest so the body is only maintaining life support functions), it is only necessary to measure oxygen consumption. An average adult at rest uses about 16 ℓ of oxygen and produces about 13.4 ℓ of CO_2 per hour. This corresponds to 80 kcal/hour or 1760 kcal/day in good agreement with other estimates of the basal metabolism rate (BMR).

It is instructive to consider the 13.4-ℓ hourly CO_2 production of an adult at rest with a view to determining the average contribution to the BMR from each type of food. Since most protein is about 16% nitrogen and most of the nitrogen is excreted in the urine, we can determine protein consumption from the nitrogen excretion rate. Typical rates are $\frac{1}{2}$ g of nitrogen per hour, which corresponds to 3.1 g of protein per hour. Since each gram of protein produces 4.3 kcal, we obtain 13.4 kcal/hour from protein. Metabolism of this protein accounts for 3 ℓ of the 16 ℓ of O_2 consumed hourly and for 2.3 of the 13.4 ℓ of CO_2 produced hourly. The remaining 13 ℓ of O_2 and 11.1 ℓ of CO_2 correspond neither to pure carbohydrate oxidation (equal amount of O_2 and CO_2) nor to pure lipid oxidation. In order to determine how much of each food type is used, we take some information from Table 2.1 and write two equations, one for CO_2 and one for O_2:

$$\text{(grams fat)} \times 1.39 \ \ell/\text{g} + \text{(grams carbohydrate)} \times 0.81 \ \ell/\text{g} =$$

$$11.1 \ \ell \ (CO_2). \qquad (2.21\text{a})$$

$$\text{(grams fat)} \times 1.96 \ \ell/\text{g} + \text{(grams carbohydrate)} \times 0.81 \ \ell/\text{g} =$$

$$13 \ \ell \ (O_2). \qquad (2.21\text{b})$$

Solving these gives 3.3 g fat per/hour and 8.0 g carbohydrate per/hour. Fat and carbohydrate each contribute about 32 kcal/hour to the BMR and we add these 64 kcal to 13 kcal/hour from protein metabolism to find 77 kcal/hour or 1800 kcal/day for the total BMR. This corresponds to an energy production of about 85 watts just to stay alive.

It should be noted that these proportions of the basic foods correspond to minimal activity. When a person is active, the muscles use carbohydrates (glycogen from blood glucose) for heavy exercise. If the demand for energy exceeds the rate at which the liver can supply glucose, then we are forced to slow down a little bit and the body begins to consume fat to produce the energy.

We now consider some of the uses of the 1800 kcal produced daily for the BMR. We inhale 360 ℓ/hour of air at room temperature and on the average, 50% relative humidity (23 mg H_2O/ℓ) and exhale it at body temperature and 100% relative humidity (46 mg/ℓ). Since 360 ℓ of air is about 16 moles, and the specific heat of air, C_p, is about 7 cal/mole-°C, it takes $(37 - 20) \times 16 \times 7 = 1.9$ kcal/hour to heat the air. In order to supply 23 mg additional water to each liter of air we must evaporate 8.3 g of water per hour and it takes 580 cal to evaporate each gram at body temperature. Therefore we use 4.8 kcal/hour to evaporate water. Finally, the hourly 360 ℓ is moved by a pressure difference of about 5 cm H_2O, which is about $\frac{1}{200}$ atm or 500 N/m^2. Since 360 ℓ is 0.36 m^3, the work PV to move the air is about 180 joules (J). Since the efficiency of muscle is only about 25%, we need 720 J/hour $= \frac{1}{6}$ kcal hr to move the air. The total contribution of our breathing is about $2 + 5 = 7$ kcal/hour or about 10% of the BMR.

Now consider the heart, which pumps about 70 cc of blood per stroke against a pressure of about 100 mm Hg. The work per stroke is therefore $PV = (7 \times 10^{-5} \ m^3) \times (\frac{100}{760}) \ (10^5 \ N/m^2) \cong 1$ J. Since there are about 72×60 strokes per hour, the heart must produce useful work of about 1 kcal/hour. Since the efficiency of most muscle is only about 25%, the heart requires about 4 kcal/hour or about 5% of the total BMR. Respiration and circulation account for about 15% of the total BMR. Comparable amounts of energy are consumed by the kidneys, digestive system, nervous system, and brain. We have already seen that keeping warm requires a very large part of the BMR energy, but some of that heat comes from the muscles of the heart and thorax (breathing) as well as from electrochemical loss across nerve fibers. Some of the heat energy is therefore used twice: once as the excess heat from useful mechanical work and again for warmth.

Part 4: Athletic Performance

It is interesting to calculate the maximum amount of work we might expect from a human being. The strongest muscles of the body are the legs, and we shall calculate how much work they can do in a day by estimating the number of flights of stairs that can be climbed assuming one climbs stairs all day long. In order not to become quickly exhausted, it is necessary to choose a slow enough pace. One flight (3 meters) in 45 seconds seems reasonable, for one flight per minute is very slow and two flights per minute is probably too fast. For an average person ($m = 70$ kg) the power output is $mgh/t = 70 \times 9.8 \times 3/45 = 47$ watts $= 42$ kcal/hour, which is considerably less than $\frac{1}{10}$ horsepower. The total amount of work that a person can do is limited by the amount of time available to produce this power—a reasonable maximum might be 10 hours/day. The total work is then about 420 kcal daily. An average, healthy person working under these conditions might need about 3500 kcal of food daily, of which about 1800 are required simply to keep alive (BMR). The remaining 1700 are used to produce the useful work of 420 kcal. These two numbers can be divided to give a muscular efficiency of about 25%.

The 25% efficiency of work from human muscles is generally accepted as a reasonable average. It has been confirmed by a variety of measurements of oxygen consumption during exercise. Some of these tests are done on a bicycle ergometer, others on a treadmill. Perhaps the most reliable measurements are done on a tiltable treadmill. The subject is asked to walk, at a fixed speed, on a horizontal treadmill for a long enough time to establish his oxygen consumption fairly well. Then the treadmill is tilted upward at some angle and the oxygen consumption is measured again. The difference between the two measurements is a good measure of the extra energy required to walk uphill and therefore do mechanical work. Measurements of this type usually yield efficiencies of about 30%.

Any system that is provided with heat or converts chemical energy to heat and then uses that heat energy to produce mechanical work is called a *heat engine*. The efficiency of heat engines has been known for a long time. The maximum attainable efficiency, e, in an ideal engine has been found to be $e = 1 - (T_c/T_h)$ where T_h and T_c are the maximum and minimum temperatures available to the engine in its cycle. For the human body, T_h is body temperature, 310°K, and T_c is ambient temperature, 293°K, from which we find $e = 5.5\%$. It is surprising that the maximum thermodynamic efficiency of the human body is found to be much lower than the observed 25%. This dilemma is not easily resolved unless we are willing to give up the idea that the body is a heat engine. The chemical energy is *not* converted into the disorderly energy of heat before it can be used to produce mechanical motion. Instead, the muscle fibers convert the energy of the chemical bonds directly into mechanical energy.

It is interesting to compare the maximum long-term energy output of a normal adult with the peak short-term output. Let us consider the act of throwing a football 60 yards (not a difficult task for someone who has practiced). Equation 1.12 gives the distance d traveled by a ballistic object when launched at a 45° angle and velocity v. We use $v = at_1 = \sqrt{gd}$ (from Eq. 1.12) and $F = Mg = ma$, which we combine to find $t_1 = 2s/\sqrt{gd}$. The launch energy $\frac{1}{2}mv^2$ divided by the acceleration time t_1 gives the average power P during the acceleration phase. Then the power

$$P = \frac{\frac{1}{2}mv^2}{t_1} = \frac{\frac{1}{2}mgd}{2s/\sqrt{gd}} = \frac{m(gd)^{3/2}}{4s}. \tag{2.22}$$

We now estimate s to be about 1 m and m to be about 1 kg, and we use $d = 60$ m. The peak power output is then 3600 watts! This is nearly 100 times larger than the average power output under conditions of heavy, steady work as calculated above. It turns out, however, that this huge instantaneous power output is rather typical of the maximum short-term output of the body under a variety of conditions.

Let us assume that the maximum throwing effort the body can produce is 3600 watts and calculate how far one can throw a 7-kg shot-put. We use the same formula and find $d = (4sP/m)^{2/3}/g = 16$ m. The Olympic record for the 16-1b shot is about 18 m and we should expect the actual throw to be a little better than our calculation because the thrower can accelerate the shot a little bit with his running and turning motion in the circle before he actually throws it. That makes the effective value of s a little larger. Increasing s allows a longer throw for a given amount of power available, and that is the reason that athletes "step into" a throw of any kind. A similar calculation using $s = 1$ m for the standing broad jump gives $d = 3.5$ m to be compared with the record of 3.3 m.

We now turn our attention to running as we did before in Chapter 1. Running at constant speed on a straight and level path should require no work since there is no acceleration and no force required. On the other hand, everyone knows that running is tiring exercise. We resolve this question by considering the movement of the runner's foot. When it is on the ground its speed is zero, but its average speed must be the same as that of the runner (it moves with the runner). We assert that the energy required for running is used to accelerate the legs and then stop them again. If a runner is traveling at velocity v and his foot is on the ground, then it has a velocity $-v$ relative to the runner. When it is swinging forward, its velocity is $+v$ relative to the runner in order to keep its average relative velocity equal to zero (it remains attached). In the runner's frame of reference, the leg is pivoted at the hip and the energy associated with its motion is $\frac{1}{2}I\omega^2$.

The runner must accelerate the angular velocity of his leg from $-\omega = -v/r$ to $+\omega = v/r$ and back to $-\omega$ for each stride (two steps). He does an amount of work $W = \frac{1}{2}I\omega^2 \times 2$ for each acceleration. We multiply this by a factor of 2 because there are two accelerations per stride, by 2 again because he must work on both legs, and by $\frac{1}{2}$ because a stride is two steps. Then we find an amount of work $W = 2I\omega^2$ per step. Using $I = \frac{1}{3}mr^2$ (for a rigid cylinder) and $v = \omega r$, we find $W = \frac{2}{3}mv^2$ for the work per step. The time t to take a step is the length of a step, s, divided by the speed. The power required to run is therefore $P = W/t = 2mv^3/3s$.* Taking the mass of a leg to be about 10 kg and the length of an average sprinting step to be about 1 m, we again use 3600 watts as a maximum power effort to calculate the maximum running speed. We find

$$v = \left(\frac{3sP}{2m}\right)^{1/3} = 8.1 \text{ m/sec} \tag{2.23}$$

which, neglecting starting time, corresponds to 100 m in 12 sec.

We now consider the running gait more carefully. The photographs in Figure 2.6 show that a runner lifts his leg high by bending his knee very

Figure 2.6 Sequence of photographs of a running man. Note the sharp bend at the knee before the leg is brought forward. (From Gray, *How Animals Move.* © 1959 Pelican Books. Reprinted by permission.)

*These more careful energy considerations tell us that the power required to run scales as v^3 and supports our conclusion in Chapter 1 that running speed is essentially independent of body size.

sharply during the part of the stride while his leg is swinging forward. This action reduces the moment of inertia substantially. If we assume that the work done when swinging the bent leg forward is negligibly small compared with the work done when swinging an extended leg, Eq. 2.23 becomes

$$v = \left(\frac{3sP}{m}\right)^{1/3}. \qquad (2.24)$$

Using $P = 3600W$ we find $v = 10.3$ m/sec and our corrected time for the 100-m dash is 9.7 sec (record = 9.9 sec).

We now return to consideration of long-term power output. A trained athlete participating in long-distance running or walking can probably put out three times as much power as the average adult who can climb a flight of stairs in 45 sec. Consider how fast an athlete should be able to walk. We use $P = 150W$ and $s = 1$ m in Eq. 2.23 and find $v = 2.8$ m/sec. The length of time required to cover 50,000 m is therefore 17,700 sec = 4.9 hours; the Olympic record for the 50,000-m walk is 4.4 hours. Long-distance running is characterized by a stride that is considerably longer than the sprint (or walk) stride. Using $s = 2$ m and $P = 150W$, we find $v = 4.5$ m/sec, which corresponds to a 10,000-m run in 37 minutes. The Olympic record for this event is about 30 minutes.

There are a number of reasons for expecting athletes to perform better than our crude approximations above indicate. For one thing, the moment of inertia of the leg was taken to be $\frac{1}{3}mr^2$, which assumes the mass of the leg is uniformly distributed along its length. It is shown in Appendix F that a more reasonable value for our tapered legs is $\frac{1}{5}mr^2$, and this correction reduces the time estimate for running events by about 18%. Table 2.2 shows the times for events we have considered in various stages of approximation along with the Olympic records. The important point is that energy considerations tell us quite a bit about the athletic performance of the human body.

TABLE 2.2   *Calculated times for common footraces*

Event	First approximation	Bent leg	Tapered leg	Olympic record
100 m	12 sec	9.7 sec	8.0* sec	9.9 sec
200 m	24 sec	19.4 sec	16 sec	19.5 sec
5000 m	22.6 min	18 min	14.7 min	13.2 min
10,000 m	45 min	37 min	30 min	30 min
50,000 m	5 hr	Not applicable	4.1 hr	4.4 hr

*A significant amount of time is spent accelerating in this event, and our failure to take it into account is responsible for part of the large discrepancy between 8.0 and 9.9 sec.

Notice that certain animals having the capability to run at high speeds for long distances have appropriately constructed legs. In particular, the running muscles of the horse and deer are found mostly in the body and upper thigh, resulting in a lower leg that is only bone and tendon. The moment of inertia is therefore considerably reduced as is the power required to run. In spite of the high sprint speed of the large cats, the horse and deer are the fastest animals in the world over any long distance.

All of these running efforts require oxygen. In order to produce power of 150 watts $= 2$ kcal/minute, we need about 0.4ℓ O_2 per minute. Since the efficiency of muscle is only about 25 %, we should need about 1.6ℓ of oxygen per minute, about three times the normal consumption and seven times basal consumption. This number is in approximate agreement with measurements of oxygen consumption rates during heavy exercise, although well-trained athletes in superb physical condition can consume up to 5ℓ/minute. In fact, maximum O_2 consumption rate is an excellent indicator of overall physical fitness.

The human athlete has an oxygen reserve of about 5ℓ that can be used to produce about 10^5 J of energy. In a sprint, there is not enough time for the body to use this extra energy source. In a 5-hour event, use of this energy only adds 0.016ℓ to the 1.6ℓ/minute consumption of the athlete. In a 5-minute event, however, use of this reserve can double the available energy and therefore play havoc with the entries in Table 2.2. Athletes are familiar with the notion of "oxygen debt" and train accordingly.

APPENDIX D
RELATIVE VELOCITY

In a collision between two equal masses where energy is conserved we find

$$v_1^2 + v_2^2 = v_1'^2 + v_2'^2 \qquad \text{(D.1)}$$

where primes denote variables after the collision. We transform to a frame of reference at rest with respect to particle 2 before the collision. Then $v_2 = 0$ and using conservation of momentum we find (see Figure D.1)

$$v_1' \sin \theta_1' + v_2' \sin \theta_2' = 0 \qquad \text{(D.2a)}$$

$$v_1' \cos \theta_1' + v_2' \cos \theta_2' = v_1. \qquad \text{(D.2b)}$$

Equations D.1 and D.2 can be combined to give

$$v_1 - v_2 = \pm(v_1' - v_2').$$

The $(+)$ sign clearly corresponds to no collision at all whereas the $(-)$ sign gives the desired result. In an energy conserving collision the relative velocity is preserved.

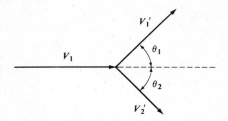

Figure D.1 Elastic collision between equal masses in a frame of reference where one of them was at rest before the collision. Note that $\theta_1 + \theta_2 = 90°$.

44

APPENDIX E
EXCHANGE DISTRIBUTION

We wish to determine the steady state temperature distribution in oppositely flowing fluid in two adjacent tubes. Consider the case of the blood supply to an animal's limb that is in an environment much colder than the core temperature of the animal, as shown in Figure E.1. The blood enters the limb

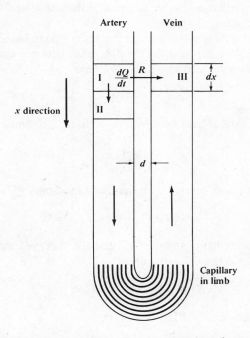

Figure E.1 Schematic diagram of countercurrent exchange loop showing regions I, II, and III for purposes of the caculation. All the energy in I must go to either II or III in the length of time it takes for the fluid to be swept out of I.

warm, flows to the extremity of the limb where it loses heat to the cold environment corresponding to a temperature drop of ΔT, and then returns in a vein that is in close thermal contact with the supply artery. We define the temperature distribution along the artery to be $T_a(x)$ and along the vein to be $T_v(x)$.

In order to simplify the calculation, we assume that both vessels have the same uniform cross-section A, that there is no heat lost to the environment anywhere except in the loop at the end of the limb (the capillary bed), and that the blood flows smoothly at constant speed v. We also assume that heat transport along the vessels is mediated only by convection (flow of blood) and between the vessels only by conduction.

The rate of conduction of heat from the warm artery to the cool vein, R_1, is

$$R_1 = [T_a(x) - T_v(x)]\frac{Ka}{d} \tag{E.1}$$

where K is the thermal conductivity of the vessel walls whose combined thickness is d and a is their contact area. For the region designated I in Figure E.1, it is clear that $a = w\,dx$ where w is the width of the contact area and dx is its length along the vessels.

Conservation of energy requires that the rate of heat flow R from region I to region III of Figure E.1 must be the difference between that flowing into region I and that flowing into region II. We have

$$R = \frac{dQ_a(x)}{dt} - \frac{dQ_a(x + \Delta x)}{dt} \tag{E.2}$$

where dQ/dt is the rate of heat flow by convection. By definition of a derivative we have

$$R = -\frac{d}{dx}\left[\frac{dQ_a(x)}{dt}\right]dx. \tag{E.3}$$

Since the amount of heat carried by a mass of blood m is $Q_a(x) = mCT_a(x)$,

$$\frac{dQ_a(x)}{dt} = CT_a(x)\frac{dm}{dt} = CT_a(x)\rho Av \tag{E.4}$$

where C is the specific heat, ρ is the fluid density, A is the cross-section area of the vessels, and v is the velocity. Then

$$R = -C\rho Av\frac{dT_a(x)}{dx}dx. \tag{E.5}$$

Now we combine Eq. E.1 and E.5 to obtain

$$\frac{dT_a(x)}{dx} = \frac{Kw}{C\rho Avd}[T_a(x) - T_v(x)] \equiv B[T_v(x) - T_a(x)]. \tag{E.6a}$$

Similar considerations for heat flow in the vein lead to

$$\frac{dT_v(x)}{dx} = B[T_v(x) - T_a(x)]. \tag{E.6b}$$

Equations E.6a and E.6b require that

$$\frac{dT_a(x)}{dx} = \frac{dT_v(x)}{dx} \tag{E.7}$$

from which

$$T_a(x) - T_v(x) = \text{constant} = \Delta T. \tag{E.8}$$

We substitute this back into Eq. E.6a and integrate to find

$$T_a(x) = T_a(0) - \Delta TBx \qquad \text{(E.9a)}$$

and

$$T_v(x) = T_v(0) - \Delta TBx \qquad \text{(E.9b)}$$

where $T_a(0) = T_v(0) + \Delta T$.

In Figure E.2 we plot the temperature distribution along the limb. We see that the temperature drop from core end to extremity is very small if ΔT is small (little heat loss), if the limb is short or the flow velocity is high (not enough time for adequate heat transfer), or if K/d is small (artery and vein are insulated from each other). It is important to note that if the limb extremity is in a severely cold environment, the loss of heat as measured by ΔT can be substantially limited by effective countercurrent heat exchange.

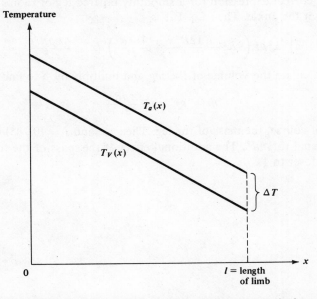

Figure E.2 Plot of the temperature (or any other parameter) along the length of the countercurrent exchange loop.

APPENDIX F
MOMENT OF INERTIA OF LEG

We calculate the moment of inertia of the human leg. For a body of arbitrary shape, the moment of inertia I is given by

$$I = \int x^2 dm \tag{F.1}$$

where dm depends on the position. Now $dm = \rho A\, dx$ where A is the cross-section area at point x that is measured along the leg from the pivot point at the hip. The human leg is about three times thicker at the thigh than at the ankle; we write $A = \pi r^2 = \pi(3R - 2xR/l)^2$ where l is the length of the leg. This is the correct expression for a smoothly tapered leg of radius $3R$ at the hip and R at the ankle. Then Eq. F.1 is

$$\int_0^l x^2 \rho \pi \left(9R^2 - \frac{12R^2 x}{l} + \frac{4x^2 R^2}{l^2} \right) dx = \frac{4\pi \rho R^2 l^3}{5} \tag{F.2}$$

We also calculate the volume of the leg and multiply by ρ to find the mass:

$$m = \rho V = \frac{13\pi \rho R^2 l}{3} \tag{F.3}$$

which is, of course, the mass of the leg. Then we find $I = (0.185)ml^2$, which is nearly equal to $\frac{1}{5}ml^2$. The additional effect of the mass of the foot brings the value closer to $\frac{1}{3}$.

Questions and Problems for Chapter 2

1. Niagara Falls is about 200 ft high. What is the temperature rise of the water, assuming all kinetic energy is converted to heat?

2. 500 g of ice ($C = 0.5$ cal/g) at $T = -20°C$ is dropped into 100 g of water ($C = 1.0$ cal/g) at $T = 10°$. What is the final temperature of the mixture (assuming no heat losses to environment)? How much ice is left? (Latent heat of ice-water transition $= 80$ cal/g.)

3. Consider two simplified animals shaped like cylinders. One has a diameter of 1 m, a length of 1 m, and a 1-cm thick layer of insulating fat. The other has a diameter of $\frac{1}{2}$ m, is 6 m long, and has a 4-cm thick layer of fat. Both have body temperatures of 35°C and live in an environment of 0°C (arctic). Which animal contains more heat? Which animal loses heat faster? In order to find which animal would survive better it is necessary to know the rate of heat loss compared with the total amount of heat contained. Which animal loses a larger fraction of its heat per unit time?

4. Two large reservoirs of heat are at 30° and 50°C, respectively, and well insulated. They are separated by a distance $3L$ and connected by a bar of square cross section with area A. This bar is made of a material whose thermal conductivity is K for two-thirds of its length and $K/2$ for the remaining one-third of its length. What is the temperature along the bar halfway between the reservoirs and what is it at the interface between the two materials? What is the rate of heat flow between the reservoirs?

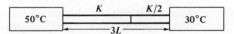

5. Consider the insulating properties of skin. Its thermal conductivity is about 4×10^{-5} cal/cm-sec-°C and its thickness is about $\frac{1}{2}$ mm. Find the heat loss for a body insulated by 1 cm of fat and $\frac{1}{2}$ mm of skin.

6. When we sleep (one-third of the day), we usually have much more insulation. Find the heat loss from the body assuming insulation of 1 cm of fat and 1 cm of clothing for two-thirds of the day and insulation of 1 cm of fat and 4 cm of blankets for one-third of the day. Ignore the skin and assume blankets act like clothing.

7. Find the skin temperature (ignoring its insulating capabilities) for 1 cm of fat and 1 cm of clothing insulation. Assume core temperature $= 37°C$ and air temperature (outside clothing) $= 20°C$. Find the hourly heat loss.

8. Suppose the ambient temperature drops from 20° to 0°C. How much extra clothing do we need to maintain the same daytime heat loss as in Exercise 7?

9. Consider a body of mass m, specific heat C, and at temperature T that is placed in thermal contact with a reservoir at $T = T_r$ via a slab of material whose thickness is d and has a thermal conductivity K. Use the heat flow equation

$Q = KAt(T - T_r)/d$ to show that the body approaches the temperature of the reservoir exponentially with time. Calculate the time constant.

10. Discuss the principle of countercurrent exchange. Show how it can work to prevent heat loss as well as to act as a multiplier. Give examples of each case.

11. Suppose I am served coffee with cream, but I'm not going to drink it for a few minutes. If I want the coffee to be as hot as possible when I drink it, should I put the cream in immediately or should I wait until just before drinking it?

12. In order to lose weight by exercising a person goes to the gym and repeatedly lifts a 15-kg mass (30 lb) from the floor to over his head (2 m). How many times must this be done to lose a pound of fat? Fat releases 9 kcal/g and the efficiency of muscle is 25%.

13. A 70-kg man does 10 chin-ups in 1 minute. (Vertical distance $= \frac{1}{2}$ m.) How much mechanical work does he do? If his muscles are 20% efficient, calculate the power required and compare it with the power of his normal BMR. The energy not used for raising his body is converted into heat that raises his temperature by ΔT. Use specific heat of body $= 1$ cal/g-°C and calculate ΔT.

14. The amount of kilocalories produced by consumption of various common foods varies over a factor of 2 but the amount of kilocalories produced by consumption of a liter of O_2 is nearly independent of the kind of food used. Why?

15. Derive Eq. 2.24.

16. Evaluate Eq. 2.22 for the 16-lb shot using $d = 60$ ft. It is a good exercise in keeping units straight.

17. Make an estimate from the photographs in Figure 2.6 of the moment of inertia of the leg during the forward swing part in a high-stepping gait. Show that the assumption leading to Eq. 2.24 is justified.

18. In order to run as illustrated in Figure 2.6 it is necessary to lift the leg. Use the photographs to estimate the amount of work done against gravity to lift the leg, and compare it with the amount of work per step to swing an outstretched leg. Show that the lifting work is small compared with the swing work.

19. Show that Eq. 1.12 and 2.22 are consistent with one another. To do this you will need some information that is not given. Estimate as well as you can.

20. The amount of mechanical work a person can do using a fixed consumption of energy (calories) is far larger than the limit imposed by the laws of thermodynamics. Explain how this is possible.

21. Assume all the work in walking is accelerating and decelerating the legs. If the moment of inertia of the leg is $\frac{1}{3}mr^2$, find the power required to walk at a speed of 1 m/sec. Use a step length of 1 m. How fast can a person walk if the power available for useful work in walking is 50 watts? How far could a person walk in an 8-hour day? Is this reasonable? Why or why not?

22. How much error is made by the approximation in Eq. 2.17?

23. Suppose that the long-term power capabilities of two athletes differ by 10% (because of, let's say, a difference in size). By how much would you expect their times for 10,000-m runs to differ?

24. Evaluate the integrals of Eq. F.2.

Bibliography

1. GEORGE BARR, *Here's Why: Science in Sports* (Scholastic Books, New York, 1962).

2. H. BLUM, Am. J. Phys. **45**, 61 (1978). (A discussion of the physics of punching and kicking in karate.)

3. L. CARLSON and ARNOLD HSIEH, *Control of Energy Exchange* (Macmillan Co., London, 1970). (A rigorous, careful treatment of a number of important problems.)

4. PAUL DAVIDOVITS, *Physics in Biology and Medicine* (Prentice-Hall, Inc., Englewood Cliffs, N.J., 1975).

5. T. RUCH and J. FULTON, *Medical Physiology and Biophysics* (W. B. Saunders Co., Philadelphia, Pa., 1960). (One of the all-time standards in physiology texts.)

6. K. SCHMIDT-NIELSEN, *How Animals Work* (Cambridge University Press, New York 1972). (A delightful description of many aspects of animal physiology. Special emphasis on countercurrent exchange.)

7. ———, "How Birds Breathe," *Scientific American* (December 1971). (A fine article on a beautiful application of countercurrent exchange.)

8. ———, *Science*, **177**, 222 (1972). (This is a superb article on animal movement, with very many references.)

9. P. F. SCHOLANDER, "The Wonderful Net," *Scientific American* (April 1957). (A superb article on countercurrent exchange.)

10. E. SELKURT, *Physiology*, (Little, Brown and Co., Boston, Mass., 1971). (A standard text.)

11. A. VANDER, J. SHERMAN, and D. LUCIANO, *Human Physiology* (McGraw-Hill, New York, 1970). (A standard text.)

12. THOMAS VAUGHAN, *Science in Sport* (Little, Brown and Co., Boston, Mass., 1970).

3

FLUIDS

Part 1: Fluid Statics

Our study of fluid statics will be concerned only with incompressible fluids (density is constant). We start with a vessel of fluid in equilibrium in a gravitional field [Figure 3.1(a)] and consider the equilibrium of a small rectangular volume of fluid with dimensions dx, dy, dz, and density ρ. There is a gravitational force dF on the volume equal to $g\,dm = g\rho\,dx\,dy\,dz$ in the downward direction and since it is in equilibrium, dF must be balanced by an upward force. This upward buoyant force arises from the pressure in the fluid; in order to support the arbitrarily placed volume element of fluid, its magnitude must also be $g\rho\,dx\,dy\,dz$ and must be independent of x, y, and z as is the gravitational dF. Furthermore, it must act upward on the horizontal surfaces of area $dy\,dx$ rather than on any of the vertical surfaces, because static fluids cannot exert a shear force. It follows that the downward force from the fluid above the volume element must be less than the upward force from the fluid below the volume element by precisely $dF = g\rho\,dx\,dy\,dz$ [see Figure 3.1(b)]. We write the pressure difference between top and bottom of the volume element $dP = (F_t - F_b)/A = -\rho g\,dz$ since the area is $dx\,dy$. We integrate this equation for dP and find $P = -\rho gz$, which is the fundamental law of hydrostatics. The pressure in any incompressible fluid in a uniform gravitational field increases linearly with depth and the $(-)$ sign shows that pressure increases in the downward direction.

52

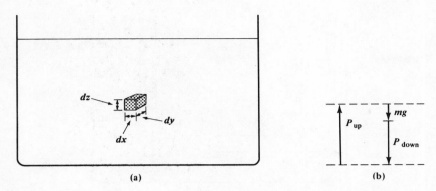

(a)

(b)

Figure 3.1 Here (a) shows a small volume of fluid at an arbitrary location that remains there because the forces on it are balanced. The force diagram in (b) illustrates the balanced forces.

There are several important consequences of this statement. First, there are no requirements on the shape of the vessel since pressure depends only on depth. This means that the pressure of the fluid at the bottom of each of the vessels shown in Figure 3.2 is the same since the fluid is of equal depth. If all the valves were opened, there would be no flow since there is no force to cause the fluid to move. It is sometimes difficult to recognize that the large amount of fluid in the left-hand vessel is supported mostly by its horizontal walls as shown by the arrows. Also note that it is possible to produce very large pressures with a tall, thin column of fluid as long as there are no other effects (such as surface tension).

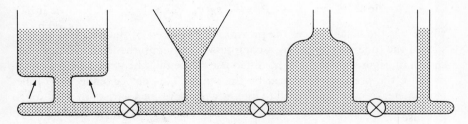

Figure 3.2 A number of differently shaped vessels connected by a common hydraulic line. The fluid level is the same in all of them.

A barometer can be constructed from a tube and a container of fluid. If a closed tube is immersed in fluid and then inverted and drawn upward as shown in Figure 3.3, the fluid will remain in the tube because there is no opportunity for air to get in and replace the fluid. Since the pressure at the surface is atmospheric, hydrostatics requires that the pressure above the

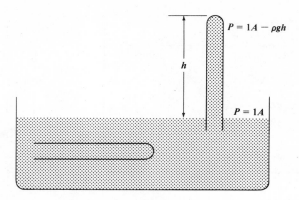

Figure 3.3 A test tube is submerged and then inverted and removed from the tray. It remains filled with fluid unless it is so tall that the pressure near the top of the column becomes zero.

surface be less than atmospheric. We write $P_h = P_A - \rho g h$ where h is the height above the surface of the fluid and P_A is atmospheric pressure. We notice that if h is large enough, say h_0, P_h becomes zero. If we raise the closed tube any higher than h_0, the fluid will not rise with it and an empty volume (nearly a vacuum) will appear. We can measure the height of the fluid in such a device (called a barometer) and use the result to determine atmospheric pressure. If the fluid is mercury, the height will average 760 mm; if it is water, the height will average about 10 m, assuming the device is at sea level.

Consider the fluid stored in the U-shaped tube of Figure 3.4(a). We have seen (fundamental law) that the pressures at any two points differ by

$$P_2 - P_1 = -\rho g (h_2 - h_1) \tag{3.1}$$

where h is the height where the pressure is measured. If the pressure on one side of the tube is changed by a compressor or pump, the fluid will move as shown in Figure 3.4(b), and the difference of the fluid heights can be used to measure the pressure produced by the pump. This device, called a U-tube manometer, is commonly used to measure pressures in a variety of situations.

The U-tube can be generalized to a hydraulic lift as shown in Figure 3.4(c). In this peculiarly shaped vessel, the pressure at A is the same as at B. If the pressure at A is increased by applying a downward force with a piston, then the pressure at B on a fixed piston will increase also. The force on a piston at B will be much larger than at A, however, because the area is larger, and such a device can be used to lift very large weights. It is important to realize though that if a weight at B is to be lifted by an amount y, the force at A must be exerted through a distance much larger than y. In fact, the displacement is just y times the ratio of the areas at A and B since the volume of fluid moved out of the thin arm must equal the volume moved into the wide

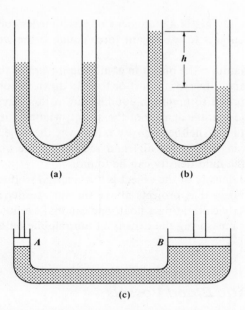

Figure 3.4 Part (a) shows a U-tube filled with fluid, with equal pressure
on each side. Part (b) shows how the fluid moves if the pressure on the right
is higher than that on the left. Part (c) shows a U-tube with unequal bores
at each end and fitted with pistons to make a hydraulic lever.

arm. The work done on both sides is the same, just as it is for a simple lever.
In fact, the quantities involved are related in exactly the same way as those
that describe the workings of a lever.

Buoyancy is a very important consequence of the fundamental law of
hydrostatics. We consider a volume of fluid $dx\,dy\,dz$ replaced by an equal
volume of any other material. The buoyant (upward) force *will not* change
because it arises from the remainder of the fluid, which is undisturbed by the
replacement. The new body, however, may not be in equilibrium because the
buoyant plus gravitational forces on it may not sum to zero (unless its average
density is the same as that of the fluid). We can write that the *net* force on a
body submersed in a fluid is

$$F = (\rho_f V - m)g = (\rho_f - \rho_0)gV \qquad (3.2)$$

where V is the volume of the object (= volume of displaced fluid), m is the
mass of the object, and ρ_f and ρ_0 are the fluid and object densities respectively.
Note that the sign of the net force (up or down) depends only on the difference
between the density of the object and the density of the fluid. If its density is
higher than the fluid's, it will sink to the bottom and rest on the floor of the
container. If its density is lower, the net force will be upward and it will rise
through the surface until the volume of fluid it displaces has the same weight

as the object itself. This is a statement of Archimedes' principle: "A submerged body is subject to a buoyant force equal to the weight of the fluid it displaces."

In order to maintain the object in equilibrium (sum of the forces be zero) it may be necessary either to support or to hold down the object. By measuring the force required to maintain equilibrium in this way, Eq. 3.2 permits us to compare the average density of the object, whatever its shape, with the density of the fluid. If neither density is known, the object density can be determined by weighing it (in air) and measuring the volume of fluid it displaces. Then the fluid density can be found.

If the average density of the object is low enough so that it will float, the amount of the object that projects above the surface depends on the fluid density. By properly calibrating a float, one can make a fluid densitometer—such a device for measuring the density of automobile battery acid is called a *hydrometer*.

Part 2: Elastic Blood Vessels

We now consider the application of fluid statics to the human body. The first and most obvious consideration stems from the variation of pressure with height. When an average individual is standing erect, his heart is about 120 cm above his feet, and his head is about 40 cm above his heart. Therefore, one expects rather large differences in blood pressure from one part of the body to another. In fact, the pressure on the blood vessels at the feet exceeds that at the heart by about $\frac{1}{8}$ atmo or the equivalent of about 125 g on each square centimeter of blood vessel. The blood vessels in the lower extremities must be capable of withstanding this additional pressure without too much stretching or they would burst.

We can discuss the behavior of fluids in elastic vessels by making a few simplifying assumptions about the walls of these vessels. Suppose we cut a section of a blood vessel, slit it, and unroll it as shown in Figure 3.5(a). We consider the piece to be a slab of uniform elastic material of width w and thickness t described by a Young's modulus Y and ask what happens when a force is applied to stretch it as shown in Figure 3.5(b). If its length was originally $2\pi r$, it will stretch by an amount (Eq. 1.27).

$$2\pi\Delta r = \frac{F(2\pi r)}{AY} \tag{3.3}$$

where $A = wt$ [Figure 3.5(b)] is the cross section and F is the applied force.

We consider what happens when the vessel is stretched by the forces derived from a pressure difference ΔP between the inside and outside. The

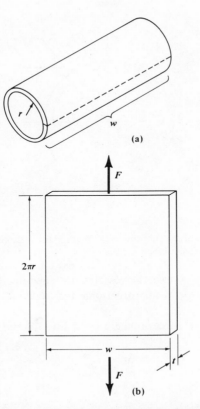

Figure 3.5 A section of blood vessel (a) which has been cut and unrolled (b).

force from the pressure acts perpendicularly to the walls and the stretch force is tangential to the walls. In order to find the proper relation between ΔP and Δr, consider a section of tubing that is cut by the vertical line as shown in Figure 3.6. The net force acting on the right half, pushing to the right, comes from the pressure difference. On any element of the wall defined by $r\,d\theta$, the horizontal component of force is $dF = \Delta P\,dA\cos\theta = \Delta P(r\,d\theta)w\cos\theta$ where w is measured along the length of the tube as in Figure 3.5. The total horizontal force F is simply the integral

$$F = \int_{-\pi/2}^{\pi/2} \Delta Prw\cos\theta\,d\theta = 2\Delta Prw. \tag{3.4}$$

This horizontal component of force acts to stretch the top and bottom of the vessel horizontally and in equilibrium is balanced by the elastic forces $F = 2AY\Delta r/r$ (from Eq. 3.3). The factor of 2 enters the elastic force because there are two sections of wall being stretched, one at the top and one at the

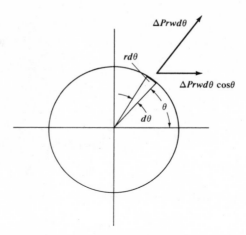

Figure 3.6 Sketch of the coordinate system used for calculating the effects of pressure in an elastic blood vessel.

bottom of the vessel. Of course, the vessel is stretched in all directions because the pressure acts equally in all directions and remains round if the walls are uniform.

We equate these two forces (elastic from Eq. 3.3 and pressure from Eq. 3.4) and find

$$\frac{\Delta r}{r^2} = \frac{\Delta P}{Yt} \tag{3.5a}$$

or

$$\frac{dr}{r^2} = \frac{dP}{Yt}. \tag{3.5b}$$

Now we integrate

$$\int_{r_0}^{r} \frac{dr}{r^2} = \int_{P_0}^{P} \frac{dP}{Yt} \tag{3.6}$$

where P_0 and r_0 are the relaxed values of P and r. Under normal circumstances, P_0 is just equal to the pressure outside the vessel when its radius is at the relaxed r_0. We perform the integration and evaluate at the limits to find

$$r = \frac{Ytr_0}{Yt - r_0(P - P_0)}. \tag{3.7}$$

Equation 3.7 diverges when the denominator goes to zero, which happens when the pressure inside is greater than the pressure outside by an amount $P - P_0 = Yt/r_0$. This pressure P is denoted P_b, the bursting pressure and is given by

$$P_b = P_o + \frac{Yt}{r_0}. \tag{3.8}$$

The consequences of the Eq. 3.8 are very curious. To begin, we notice that P_b is considerably larger for small vessels than for large vessels that have the same wall thickness. In order for P_b to be the same for different size vessels, the wall thickness must scale with the vessel radius. Expressed another way, a smaller vessel can withstand a greater pressure than a larger one with the same wall thickness. Note that Eq. 3.7 describes how the radius r of the vessel increases with pressure, and as we shall see, flow rate depends on r^4. We therefore expect the volume flow in elastic vessels to increase more rapidly than ΔP, the applied pressure. Thus if a pressure pulse such as a heartbeat is applied, elastic vessels carry fluid more readily than rigid ones. We also note that the fractional increase in radius $dr/r = r\, dP/Yt$, increases with r and the fractional increase in volume $dV/V = 2r\, dP/Yt$ does likewise. Just as in the case of the soap bubble (see surface tension), a given fractional increase in pressure begets a much larger fractional increase in volume when the radius is larger than when it is small. Small vessels are "tighter" than large ones. (We also note that dV/dP is proportional to r^3.)

P_b depends on Y, the Young's modulus of the vessel walls. There are some diseases, notably syphilis, that weaken the arterial walls by causing a decrease in Y. The result is that the victims are likely to suffer massive arterial failure; if it occurs in the brain, the victim can die of a stroke.

Figure 3.7 shows how the radius changes with a change in pressure. Notice that if P increases, so does $P - P_0$ and the quantity $Yt - r_0(P - P_0)$ decreases causing an increase in r. There is always a new equilibrium value of r as long as P is less than P_b. Similarly, a decrease in P leads to another

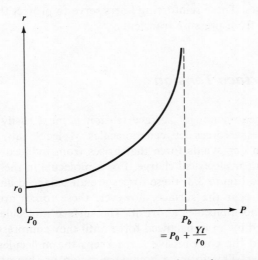

Figure 3.7 Plot of blood vessel radius versus internal pressure (Eq. 3.7). Note that at $P = P_b$ the radius goes to infinity and the vessel bursts.

equilibrium at a reduced value of *r*. The vessel is stable against small displacements from equilibrium.

We now consider the special case where the vessel is not stable against
small displacements from equilibrium because the walls of the vessel are
restricted by a constant tangential force F_o (such as a muscle force) rather
than by an elastic force that increases with an increase in size. Of course,
Eq. 3.4 still applies, and we choose $\Delta P = P_i - P_e$ where P_i and P_e are the
interior and exterior pressures. We find

$$P_i = P_e + \frac{F_0}{2rw} \tag{3.9}$$

which means that, if a vessel is constricted momentarily in some place, the
pressure in it will increase causing the blood to be forced out of that section.
The vessel will then collapse and the blood will stop flowing until there is
enough of a build-up of pressure upstream of the constriction to force the
vessel open again. This process of collapsing and opening is self-perpetuating:
the vessel flutters. The instability derives from the presence of constant
forces rather than elastic ones that become very small as the vessel starts to
collapse.

Arteries are protected against the fluttering by the presence of elastic
material in their walls (elastin). They are also protected against bursting by
the presence of tough fibers of collagen coiled in the walls. As the walls are
stretched under the increase in pressure applied by the heart, these fibers
uncoil until they are fully extended. Since they will not stretch easily, they
maintain the radius of the vessel nearly constant thereafter even as the
pressure increases. These reinforcing fibers serve to protect the large arteries
against rupture from pressure transients.

Part 3: Surface Tension

The familiar phenomenon of surface tension is most easily understood in
terms of the cohesive forces between molecules. Molecules attract one another
through the van der Waals force that arises from induced changes in the
distribution of their electrical charge. For a molecule in the midst of a stationary fluid (see Figure 3.8) these forces are all balanced and sum to zero.
For a molecule near the surface, however, these forces are not balanced
since there are no molecules above it. The surface molecules are therefore
pulled downward by an unbalanced force until they compress the molecules
below them, and the compressive force keeps the molecules from moving
down any further. This forms a boundary region or skin at the surface of

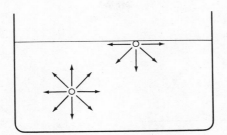

Figure 3.8 Schematic diagram of the forces on a molecule in the bulk of a liquid and at its surface.

the fluid where the density is higher and the forces between the molecules are different from those in the bulk material. In order to move a molecule from the bulk to the surface, and thereby increase the surface area, work must be done. This work is done against the force we call surface tension.

The magnitude of the force can be determined with the device shown in Figure 3.9. The shaded bar can slide to the right, but in order to do so it must enlarge the surface area of the liquid and thereby bring more molecules to the surface. This requires work and the work is the force times the distance moved by the bar: $W = Fx$. Since the film of liquid has two surfaces, the force is given by $F = 2a\gamma$ and the amount of work by $W = 2a\gamma x$. γ is called the surface tension and has dimensions force/length or energy/area, and a is the length of the bar.

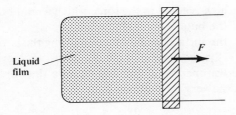

Liquid film

F

Figure 3.9 Sketch of one kind of device for measuring surface tension.

We now consider a spherical bubble that remains inflated by having a higher pressure inside than the ambient outside pressure. The pressure difference is ΔP. We will derive an expression relating ΔP and r to γ, the surface tension. First we show that the effect of ΔP on a hemisphere is a force $\Delta P(\pi r^2)$ directed parallel to the axis of symmetry of the hemisphere. In Figure 3.10(a), we see that the area element dA on the surface has the magnitude

$$dA = (r \sin \theta \, d\phi)(r \, d\theta) = r^2 \sin \theta \, d\theta \, d\phi \qquad (3.10)$$

where the coordinates are defined in Figure 3.10(b). Next we calculate the

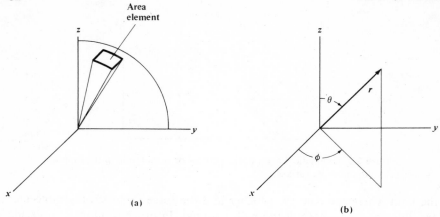

Figure 3.10 Coordinate system on the surface of a spherical bubble. In (a) the illustrated element of area has dimensions $r \sin \theta \, d\phi$ in one direction and $r \, d\theta$ in the other. The usual spherical coordinate system is shown in (b).

total z-component of force on one hemisphere. We write

$$dF_z = \Delta P \, dA \cos \theta = \Delta P r^2 \sin \theta \cos \theta \, d\theta \, d\phi \qquad (3.11)$$

and integrate to obtain

$$F_z = \int dF_z = \Delta P r^2 \int_0^{2\pi} d\phi \int_0^{\pi/2} \sin \theta \cos \theta \, d\theta = \Delta P r^2 (2\pi)(\tfrac{1}{2}). \qquad (3.12)$$

The force is $\Delta P(\pi r^2)$, which is equal to the force which would be produced on an equatorial plane of the bubble.

On the other hand, the surface tension force holding the two hemispheres together is $F = 2\gamma L = 2\gamma(2\pi r)$ where the first factor of 2 arises from the existance of two surfaces, one inside and one outside the bubble. Equating these two forces gives the very interesting result

$$r = \frac{4\gamma}{\Delta P}. \qquad (3.13)$$

Equation 3.13 is interesting because it shows that as the pressure rises in a bubble, the radius becomes smaller. If we pump a large amount of air in, the bubble expands so much that the pressure drops substantially. We have all experienced this consequence when we inflate a balloon: it's hard to get started, but once the balloon has reached a certain size, it's much easier to keep blowing air in. Equation 3.13 can be written $\Delta P = 4\gamma/r$, which shows that the pressure is higher in a smaller bubble. This means that if two bubbles are mounted on the end of a tube as shown in Figure 3.11 and the valve is opened, the small one will shrink, injecting all its air into the larger one. The volume increase of the larger one will be greater than the original

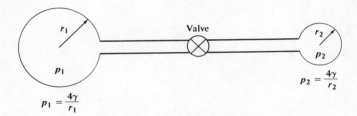

Figure 3.11 If two bubbles of the same material of unequal size are connected, the big one will grow at the expense of the small one because the pressure in the small one is higher. The volume of the big one increases by more than the volume of the small one decreases.

volume of the smaller one since the pressure is lower. The implications for aneurysm* or other problems are obvious.

The human lung permits the exchange of gases with the blood by bringing it into contact with the air inhaled into a large number of tiny air sacs called *alveoli*. These alveoli are approximately 0.1 mm diameter when collapsed, so that, assuming the forces are not elastic but arise from surface tension, the pressure required to inflate them is $\Delta P = 2\gamma/r = 2(50)/0.01 = 10^4$ dynes/cm^2. Here we have removed the factor of 2 because there is only one surface (the alveoli are essentially air-filled cavities in the lung tissue and therefore have only one fluid-air surface) and have taken $\gamma = 50$ dyne/cm, which is the surface tension of water. This pressure is about $\frac{1}{100}$ atm or about 8 mm Hg = 4 in. H$_2$O. Since the area of the average adult diaphragm is about 500 cm^2, the force required to breathe is about 50 N, the weight of 5 kg. A newborn infant has a considerably smaller diaphragm but still must exert a force of many newtons in order to breathe. In order to facilitate breathing, the alveoli are normally coated with a *surfactant* (surface tension reducing agent) to make breathing easier. If the surfactant is missing, the consequence is called hyaline membrane disease and is usually fatal.

One of the most common manifestations of surface tension is capillary action. When a thin tube is inserted into a liquid, the liquid is often drawn up into the tube as shown in Figure 3.12. We can easily calculate the height to which the fluid rises as follows. The vertical component of the surface tension force is $\gamma \cos \theta$ per unit length of interface and the length of the interface is $2\pi r$ where r is the radius of the tube. The total force of surface tension is $2\pi r\gamma \cos \theta$ and this is equated to the weight of the raised fluid $\rho\pi r^2 hg$. Then $h = 2\gamma \cos \theta/\rho g r$. The angle θ is a constant of the fluid-tube surface and does not depend on h.

---

*Aneurysm is the inordinately large swelling that occurs in pressurized vessels when their walls are too weak to contain the fluid. The walls can have elastic tension (such as arteries) or constant tension (muscle or surface).

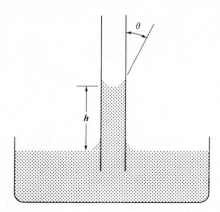

Figure 3.12 Capillary action in a thin tube. θ is called the *wetting angle*.

We might inquire about the highest point to which water can rise in a capillary tube. Clearly the pressure in the fluid cannot be negative at the top of the column (point of lowest pressure) and hydrostatics therefore limits the height of the column to $h = \Delta P/\rho g = 10$ m where $\Delta P = 1$ atm. In order to raise the fluid 10 m, the radius of the tube must be $r = 2\gamma \cos\theta/\rho gh = 10^{-4}$ $\cos\theta$ cm $= \cos\theta$ microns. Since $\cos\theta$ is always less than unity, it is clear that the tube will be less than a micron radius.

The question of how sap rises in trees to heights much greater than 10 m (some trees are 100 m tall) is not easily answered in terms of surface tension and hydrostatics. It is necessary to consider the cohesive forces (tensile strength) that determine the properties of the fluid. When the sap is considered to have some tensile strength, hydrostatics does not apply and the pressure drop is no longer simply ρgh. The column of sap is pulled up through tubes (xylem tubes) in the tree trunk by forces that result from evaporation of water from the leaves. Nutrients are thus transported to the highest parts of the tree.

Part 4:   The Flow of Frictionless Fluids

We shall study the steady flow of incompressible fluids. By *steady* we mean that the velocity of the fluid at every point in space can be described by a vector field that does not change in time. Also, we deal only with nonturbulent, smooth flow, called *laminar flow*, that is illustrated in Figure 3.13. The lines of flow are called *streamlines* and represent the path of a microscopic element of fluid in the flow. Since the velocity field remains unchanged in time, every element of fluid on a particular path or streamline remains on it.

Figure 3.13 Streamlines in a moving fluid.

Furthermore, any element of fluid to be found anywhere on a particular streamline must have started on it. It is clear from the above that streamlines never cross and fluid does not cross streamlines in laminar flow.

We consider a group of streamlines that are arranged so that they form an imaginary closed "pipe" through which the fluid flows. Of course, there is no actual boundary, but there might as well be since the fluid does not cross streamlines. Consider what happens to an element of fluid of mass m that fills a section of this pipe as it flows along (Figure 3.14). Changes in its velocity v or its height h will result in changes in its kinetic energy $\frac{1}{2}mv^2$ or its potential energy mgh. These changes are brought about by work done by the forces that derive from the pressure of the fluid acting at the boundaries of the volume element of fluid in the pipe. As a volume element moves along

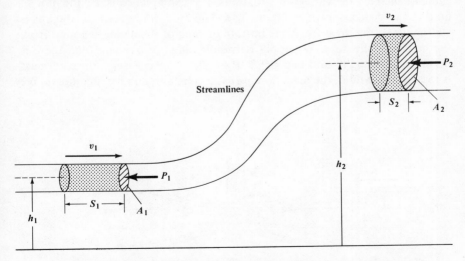

Figure 3.14 A virtual pipeline defined by a surface of streamlines in a moving fluid. No real pipe exists, but since fluid does not cross streamlines, there might as well be a surface. The volumes $A_1 s_1$ and $A_2 s_2$ are equal but travel at different speeds.

through the imaginary pipe, the work done by the force moving it to the right is just balanced by the work done by the force opposing that motion, unless the volume changes shape (we neglect friction for now). Then the difference is the amount of work $A_1 P_1 s_1$ at the left minus $A_2 P_2 s_2$ at the right. We can use energy conservation and write the final energy = initial energy + work done as follows:

$$\tfrac{1}{2} m v_2^2 + m g y_2 = \tfrac{1}{2} m v_1^2 + m g y_1 + (A_1 P_1 s_1 - A_2 P_2 s_2). \qquad (3.14)$$

Note that $A_2 s_2 = A_1 s_1$ = volume of fluid element (it doesn't change because the fluid is incompressible). We divide both sides of this equation by the volume and find the quantity $\tfrac{1}{2} \rho v^2 + \rho g h + P$ is a constant of the motion that is unchanged by moving from (1) to (2). Since any set of streamlines can be chosen, this argument generally describes the flow of fluid through a pipe. The statement

$$\tfrac{1}{2} \rho v^2 + \rho g h + P = \text{constant} \qquad (3.15)$$

is called Bernoulli's law. Although it rarely applies exactly (we have neglected friction), it is very useful in a variety of situations involving the description of moving fluids. Notice that for the special case of a static fluid ($v = 0$) it is equivalent to the hydrostatic equation (Eq. 3.1). In the case of viscous flow, we can subtract the work done by the viscous forces from the energy equation above.

Bernoulli's law is extremely important because it is approximately applicable to such a wide variety of situations. For example, consider the flow of fluid in the system shown in Figure 3.15. The fluid in the vertical columns is stationary so the pressure at the bottom of each of them can be found from the height of the fluid. We apply Bernoulli's law, recognizing that there is no change in height, and find $\Delta P = P_1 - P_2 = \tfrac{1}{2}\rho(v_2^2 - v_1^2)$. Since no mass is lost from the fluid, the mass flow $\rho A v$ is a constant, and for incompressible

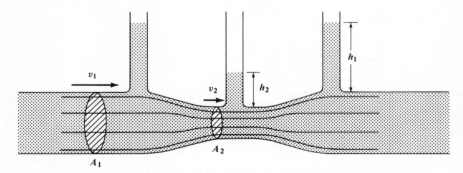

Figure 3.15 Fluid flow in a constricted tube. Because the flow is faster where the tube is narrow, the pressure is lower. Therefore, the column of standing fluid is shorter.

fluids $\rho = $ constant. Therefore, we have $A_1v_1 = A_2v_2$ or $v_1 = v_2(A_2/A_1)$, which gives

$$\Delta P = P_1 - P_2 = \frac{1}{2}\rho v_2^2\left[1 - \left(\frac{A_2}{A_1}\right)^2\right]. \tag{3.16}$$

If the cross-section areas of the pipe are known and ΔP is measured using the heights of the fluid columns (remember $\Delta P = \rho gh$), one can find the velocity v_2 and, therefore, the volume flow rate $A_2v_2 = A_1A_2\sqrt{2g\Delta h/(A_1^2 - A_2^2)}$. Such an arrangement can, therefore, be used to measure (and regulate) flow. It is called a Venturi flowmeter.

Another application of Bernoulli's law is in the description of the aspirator. Water is forced under high pressure through a constriction (called a Venturi) that has a port at the narrowest portion (Figure 3.16). If the cross section at that point is small enough, the pressure in the tube leading to the port will be below one atmosphere and a partial vacuum is developed.

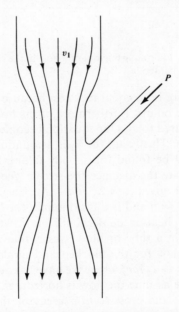

Figure 3.16 Fluid flow in an aspirator. P can be less than one atmosphere if v_1 is large enough.

Several other phenomena can be described using Bernoulli's law including the operation of airplane air speed indicators and the flight of a curve ball in baseball or a slice in golf. On the other hand, airplane flight depends to some extent on the compressibility of air, although Bernoulli's law is often used to describe the origin of the lift force on the wing.

Consider the following problem. A container of water is filled to a depth d with a liquid and there is a hole at the height h above the bottom of the container (Figure 3.17) What is the speed of the water that flows out through the hole and how far does it go? We consider a streamline that intersects the surface of the water at the top, follows some path through the liquid, and goes through the hole. We write Bernoulli's law on this streamline for a point on the upper surface and one near the exiting liquid:

$$P_1 + \rho g d + \tfrac{1}{2}\rho\, v_1^2 = P_2 + \rho g h + \tfrac{1}{2}\rho\, v_2^2. \tag{3.17}$$

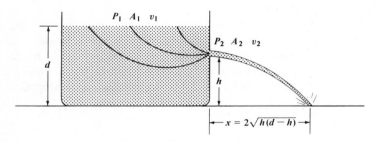

Figure 3.17 Fluid squirting out of a hole in the side of a filled vessel. The trajectory is parabolic.

Now P_1 and P_2 are both equal to 1 atm because the fluid surface is exposed to the air at those points. Furthermore, if the hole is fairly small, v_1 is negligibly small compared to v_2. Our equation becomes $2\rho g(d - h) = \rho v_2^2$ whence $v_2 = \sqrt{2g(d - h)}$. This result for v_2 is astounding because it is the same as that which would be found for a body falling freely through the distance $d - h$. We calculate the distance the stream goes x by using the flight time $t = \sqrt{2h/g}$, which gives $x = 2\sqrt{h(d - h)}$. Clearly the distance is zero for both $h = 0$ and $h = d$ and is a maximum d for $h = d/2$.

In another application of Bernoulli's law, we consider what happens to the flow of fluid in a tube or blood vessel that has a partial blockage. We write Bernoulli's law for the blocked and unblocked regions of the vessel $P_1 + (\tfrac{1}{2})\rho v_1^2 = P_2 + (\tfrac{1}{2})\rho v_2^2$ where we have left out the gravitational potential energy because we assume the flow is horizontal. We note that $v_1 A_1 = v_2 A_2$ where the A's are the cross-section areas of the vessel and find $P_2 = P_1 - (\tfrac{1}{2})\rho v_2^2[1 - (A_2/A_1)^2]$ as in Eq. 3.16. If the blockage is at (2) so that A_2 is less than A_1, it is clear that there can be a substantial reduction in pressure at the blockage. When the static pressure P_1 is adequately high as it is in the arteries, there will be little consequences, but if it is low, as in the veins, the total pressure P_2 could become zero and the vein will collapse. The blood will stop flowing and then the pressure will start to build up until it once again opens the vein. When the blood has been accelerated to high enough speed, the pressure will drop again and the vein will collapse again. This fluttering of the veins can often be heard through a stethoscope and is often the key to diagnosis of vascular disease.

A *catheter* is a thin tube that can be inserted into a vein or artery from the outside (usually through a large hypodermic needle). Catheters can carry a saline solution similar to blood and can be connected to a manometer. Since the fluid in it does not flow with the bloodstream, the manometer can provide a direct measure of the pressure inside a blood vessel. By moving the catheter through the vessel and watching the manometer, a diagnostician can detect sudden pressure changes and determine the location and size of vascular constrictions.

One must be very careful to distinguish between kinds of catheters with either open ends or holes in the sides. In Figure 3.18 we see schematic diagrams of a catheter with both types of openings. At point 2 the velocity is very much smaller than it is at 1 or 3, and so the pressure measured by an end-open catheter would be considerably higher than the actual pressure.

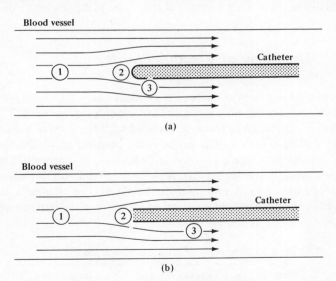

Figure 3.18 Streamlines of flow in a blood vessel containing a catheter. Part (a) illustrates an open-side and (b) an open-end catheter.

We find $P_2 = P_1 + (\frac{1}{2})\rho v_1^2$. Note that the pressure P_3 is much more nearly equal to P_1 as long as the diameter of the catheter is small compared with the diameter of the blood vessel. We have $P_3 = P_1 - \frac{1}{2}\rho v_1^2[a^2(2R^2 - a^2)/(R^2 - a^2)^2]$ where R is the vessel radius and a is the catheter radius. When a is small compared with R, the quantity in brackets is much less than unity and we see that the flow correction to the measured pressure using a catheter with a side opening is much smaller than it is using a catheter with an end opening.

We now consider how to determine the effectiveness of a heart valve using the pressure reading at different points in the aorta (see Chapter 4 page 81 for anatomy). Bernoulli's equation gives the pressure difference between the vessel and the ventricle as $\Delta P = \frac{1}{2}(v_1^2 - v_2^2)\rho$. If we assume the velocity in the ventricle is negligibly small compared with the velocity in the aorta, and also note that the cardiac output or volume flow dV/dt is Av where A is the cross section, we find $2\Delta PA^2 = \rho(dV/dt)^2$ from which the effective area of the heart valve can be calculated if the cardiac output is known.

It should be pointed out that, even in the absence of any friction, the flowing fluid produces a force on the end of the catheter (or on any object) that is $\frac{1}{2}\rho Av^2$. This force arises because the layer of molecules along the surface of the object has zero velocity and it is called the *pressure force* or the *inertial force*. It provides the basis of the v^2 term of fluid friction. Fluid friction is discussed in Part 6 of this chapter and in Appendix C, p. 19.

Part 5 :   The Flow of Viscous Fluids

Up to now we have been concerned with ideal fluids that flow without friction (viscosity $= 0$). We now consider the flow of a viscous fluid. The viscous force arises from the shear forces associated with the flow of a thin sheet of fluid past a surface with a different velocity, possibly zero. The shear force F is $F = \eta A(dv/dy)$ where η (lowercase Greek eta) is a proportionality constant called *viscosity* (see Table 3.1), A is the area common to the surface and the sheet of flowing fluid, and dv/dy is the sheet. The change in velocity as one moves in the perpendicular direction away from the surface whose area is A.

TABLE 3.1   *Viscosities of several common fluids*

Fluid	Viscosity (Poise)*	Temperature (°C)
Water	0.010	20
Water	0.0070	37
Blood	0.040	37
Blood plasma	0.015	37
Air	1.8×10^{-4}	20
Glycerin	15	20
Mercury	0.015	20

*The cgs dimension of viscosity is dyne-sec/cm² and is called a Poise. Viscosities are often given in centiPoise = 0.01 Poise.

Let us study the case of laminar flow of an incompressible, uniform fluid in a round pipe with rigid, smooth walls. (Naturally, this case is very idealistic, but it should provide a model from which we can work toward more realistic situations.) A simple symmetry argument demonstrates that the fluid velocity can only vary with the distance from the center of the pipe r, and the net force dF on a particular cylindrical shell (difference between forces from inner and outer shells) of radius r and length L is

$$dF = F(r) - F(r + dr) = \eta\left[2\pi rL\frac{dv}{dr} - \underbrace{2\pi L(r + dr)}_{\substack{A \text{ at} \\ r = r + dr}}\underbrace{\left(\frac{dv}{dr} + \frac{d^2v}{dr^2}\,dr\right)}_{\substack{\frac{dv}{dr} \text{ evaluated at} \\ r = r + dr}}\right]$$

$$= -2\pi\eta L\,dr\left(\frac{dv}{dr} + r\frac{d^2v}{dr^2}\right) \equiv -2\pi\eta L\,dr\,\frac{d}{dr}\left(r\frac{dv}{dr}\right). \tag{3.18}$$

The force needed to keep this fluid flowing steadily must just balance this viscous force. The required force is the product of the pressure difference ΔP and the end area of a cylindrical shell or $dF = \Delta P\,dA = \Delta P(2\pi r\,dr)$. We equate these forces and find

$$-r\Delta P = \eta L\frac{d}{dr}\left(r\frac{dv}{dr}\right). \tag{3.19}$$

This can be readily integrated with respect to r to give

$$-\frac{\Delta Pr^2}{2\eta L} = r\frac{dv}{dr} + C \tag{3.20a}$$

which gives

$$\frac{dv}{dr} = \frac{\Delta Pr}{2\eta L} - \frac{C}{r}. \tag{3.20b}$$

We integrate this again with respect to r and find

$$v = -\frac{\Delta Pr^2}{4\eta L} - C\ln r + C'. \tag{3.21}$$

Clearly C must be zero or else v would be infinite at the center of the pipe ($r = 0$). Also, we choose C' so that $v = 0$ at the walls of the pipe ($r = R$). Then we find

$$v = \frac{\Delta P}{4\eta L}(R^2 - r^2) \tag{3.22}$$

which is a description of how the fluid velocity varies with distance from the center of the pipe. The velocity distribution is parabolic with radius and is zero at the walls. The maximum velocity occurs at the center and depends on the pressure applied to push the fluid through the pipe. The flow can be imagined as a series of cylindrical shells moving with different velocities through the pipe.

The total volume flowing in the pipe, dV/dt, can now be calculated from this parabolic velocity distribution. For any particular cylindrical shell of area $dA = 2\pi r\, dr$, the flow is

$$v\, dA = \frac{\Delta P}{4\eta L}(R^2 - r^2)(2\pi r\, dr) \tag{3.23}$$

and this can be integrated from $r = 0$ to $r = R$. The result is called Poiseuille's law:

$$\frac{dV}{dt} = \frac{\pi R^4 \Delta P}{8\eta L}. \tag{3.24}$$

We must always remember that Poiseuille's law is not a complete or an accurate description of blood flow. All three major assumptions in the derivation of Eq. 3.24 are violated in one or more parts of the circulatory system:

1. The fluid is not homogeneous—it contains particles such as red and white cells whose size is large enough to affect the flow in small vessels such as arterioles, venules, and capillaries.
2. The flow is not steady—it is driven by the pulsating heart resulting in pressure fluctuations and shock waves that affect flow in arteries.
3. The walls of the vessels are neither straight nor rigid—they are partially elastic and their compliance (stretchiness) is not constant. The wall movement affects flow in both arteries and veins.

Nevertheless, Poiseuille's law is used to describe a wide variety of situations where there is laminar flow. It predicts that the flow rate is proportional to R^4, which means a reduction of less than 20% in the radius of a tube will cut the flow in half or will require double the pressure to maintain the same flow.

When fluid flows through a constriction that is small enough so that the speed of the fluid in the constriction is much larger than the speed outside of it, Bernoulli's equation becomes $\Delta P = (\frac{1}{2})\rho v^2$ and since $vA = dV/dt$, $dV/dt = \sqrt{2A^2\Delta P/\rho}$. Notice that the flow in this case is proportional to $\sqrt{\Delta P}$ rather than to ΔP as in the case of viscous flow.

Do not be misled by this apparent contradiction with Eq. 3.24. In one case ΔP refers to the pressure difference at either end of a straight, uniform tube and in the other case ΔP refers to the pressure difference between the point where the fluid is flowing slowly in the wide part of the tube and the point where it is rushing through the narrow part. Bernoulli's law neglects viscous friction and the pressure differences in Bernoulli flow arise from inertial effects.

It is often helpful to make an analogy between fluid flow and electric current flow. We allow the fluid volume to be analogous to charge and the pressure that causes the fluid to flow to be analogous to voltage. Then

Poiseuille's law is analogous to Ohm's law where the resistance Z is given by

$$Z = \frac{8\eta L}{\pi R^4}.$$ (3.25)

We can readily derive rules for the resistance of tubes in parallel or in series and show that they are the same rules for electric circuits. Furthermore, since the resistance of a single vessel depends on the fourth power of its radius, it's clear that small changes in the radii of peripheral vessels can provide for a sensitive regulation of blood flow. The total peripheral resistance (TPR) can be found from the flow and pressure. Small dimensional changes are sufficient to direct large quantities of blood to particular areas of the body such as the digestive system (after a big meal) or to the skin (when overheated).

We can carry the electrical analogy further by considering the elastic properties of the blood vessels. They store blood under pressure just as a capacitor stores charge with an applied voltage. The capacitance is not constant, however, but is found from Eq. 3.7. The volume of a length L of blood vessel is

$$V = \pi r^2 L = \frac{\pi Y^2 t^2 L}{(P_b - P)^2}$$ (3.26)

where P_b, the bursting pressure, is given by Eq. 3.8. The fluid analog of electrical capacitance $C = $ charge/voltage becomes

$$C = \frac{\text{volume}}{\text{pressure}} = \frac{\pi Y^2 t^2 L}{(P_b - P)^3}$$ (3.27)

for the elastic vessel. Obviously this depends on P and therefore cannot be treated exactly analogously to an electrical capacitor.

Nevertheless, we can use our electrical intuition to understand certain phenomena. For example, a blood vessel can be thought of as a distributed capacitance and resistance with a characteristic RC time constant. Its reduced ac response substantially reduces the amplitude of the pulses from the heartbeat, and eventually produces a steady flow of blood in the capillaries. One can even calculate the time constant, phase shifts, and dispersion relations of this "transmission line" by associating the kinetic energy of the moving blood with inductive reactance.

Part 6: Fluid Friction

The warning at the start of Appendix C on p. 19 is worth repeating: The problem of fluid friction has not been solved for the general case. There are partial or approximate solutions to many special cases, and some of these can be extended beyond their original domain. For example, in some cases the change of density of gases is so small that the solutions for incompressible fluids may be applied. It is usually necessary, however, to solve each fluid friction problem for its particular application.

Let us consider the problem of scaling in fluid friction. Suppose that a sphere of radius r moves through a fluid of density ρ and viscosity η at velocity v and that it is subject to a friction force F as a result of this motion. We ask how this force changes as a function of density, viscosity, size, and speed. In the 1880's, Osborne Reynolds found that the dimensionless ratio of inertial to viscous forces, which is now called Reynolds' number, must be maintained constant as different quantities are varied in order to scale fluid friction problems.

The inertial force is found from Bernoulli's law as $\frac{1}{2}\rho A v^2$ where A is the area of the object transverse to the flow. The viscous force is given by ηA (dv/dy) where A is the area over which the fluid flows. It is the area "parallel" to the flow which determines the viscous force and the area "perpendicular" to the flow which determines the inertial force. For the special case of a sphere, these two areas differ by a factor of 2. The quantity (dv/dy) is a measure of how far out into the fluid the influence of the moving sphere is felt. If we approximate it by the ratio of the velocity v to some dimension in the fluid d, we find that Reynolds' number is

$$N_R = \frac{\rho v d}{\eta}. \tag{3.28}$$

The Reynolds number determines the dynamic character of fluid friction problems. It is possible to model real problems in a laboratory environment (wind tunnel, towing tank) only if N_R in the model is the same as it is in the real situation to be studied.

Motion through a fluid at very low Reynolds number is governed primarily by viscous forces. In this case the friction force is proportional to velocity. For the case of a sphere, the friction F is given by Stokes' law

$$F = -6\pi\eta r v \tag{3.29}$$

where r is the radius of the sphere. In this case the measure of distance in the fluid where the moving sphere exerts its influence is approximately the same as the sphere's diameter. We note that this expression, which is only valid for small N_R, does not depend on the fluid density. Experiments with gases (viscosity only weakly dependent on pressure) show that the terminal velocity for small objects is essentially unchanged when the pressure (density) is varied over a factor of 100. This is because the forces required to push the gas molecules out of the way of the moving body are negligible compared to the viscous forces in determining the friction.

We can look at this motion another way. In Figure 3.19(a) we see the streamlines around a sphere moving in a fluid. The viscous force arises from the motion along the surface from point A to B past C or C'. The pressure of

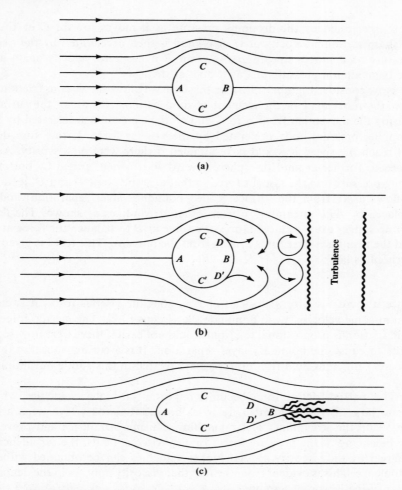

Figure 3.19 Part (a) shows laminar flow (low N_R) around a sphere. In (b) the speed is large enough for the development of turbulence. In (c) the speed is the same as in (b) but the streamlined shape reduces the friction force. The force reduction occurs because the turbulence affects a smaller area.

the fluid is the same at both A and B and there is no resulting pressure force (inertial force). The speed of the fluid at C and C' is larger as result of the flow, and consequently the pressure is lower in accordance with Bernoulli's law, but this pressure acts perpendicular to the direction of motion and doesn't produce a friction force. It is easy to see that the pressure forces at A and B are balanced and that there is no net pressure force on the sphere.

Let us consider the motion of a small element of fluid along the surface of the sphere. At point A it is brought to a stop relative to the surface and then accelerated by the dropping pressure as it moves toward C or C'. It reaches maximum velocity at C or C' and is then decelerated by the rising pressure as it moves toward B. At point B it has zero velocity again and exerts maximum pressure just as it does at A.

Now consider Figure 3.19(b). In this case we allow the viscous friction to slow the movement of the element of fluid as it moves along the surface. During the movement from A to C and C' the element is accelerated by the dropping pressure and decelerated by the viscous force. It therefore does not reach the speed it would have attained if there were no viscosity. As it proceeds further around the sphere toward B, it is decelerated by both the rising pressure and the viscous force so that at some point D and D' it stalls. It flows away from the surface (called boundary layer separation) and is replaced by fluid streaming in from other streamlines as shown. The fluid can no longer exert as large a pressure force at B to balance the force at A, and the result is a drag force arising from the pressure. This force is given by Bernoulli's law

$$F = \tfrac{1}{2}\rho A v^2. \tag{3.30}$$

Since it increases with v^2 instead of v as does the viscous force, it is clear that at some velocity it will dominate the viscous force. The inertial force of fluid friction is proportional to v^2 but would not exist if there were no viscous force to cause the boundary layer separation. It is a curious fact that these two very different fluid friction forces are related in this subtle but intimate way.

The inertial friction force becomes important at higher velocities, which means large Reynolds number. It arises because boundary layer separation breaks up the streamlines of laminar flow causing turbulence and prevents the balancing of the pressure force. In very viscous fluids, the turbulence is damped by the viscosity so that higher velocities can be obtained without turbulence. This is evident from the fact that viscosity appears in the denominator of the Reynolds number.

The foregoing discussion should provide us with a physical basis for understanding the effect of streamlining. In Figure 3.19(c) we see that a streamlined object can be shaped so that the pressure rise from C and C' to B is very gradual. This results in a minimal deceleration of the fluid moving in the boundary layer so that the point of boundary layer separation D and D' is pushed far to the back of the object. The pressure force at A is balanced by the forward component of the pressure force along the entire length from C to D resulting in a larger cancellation effect and therefore less drag. The region of turbulence is much smaller as a result. Streamlining therefore reduces the fluid friction for motion at high Reynolds number.

On the other hand, we might expect streamlining to increase the effects of viscous friction because there is a larger area for the viscous force to act upon. This expectation is supported by experiment. In viscous fluids or at low velocities (both of which produce small N_R) streamlining is a hindrance to motion, not an aid. For the commonly encountered cases of automobiles, birds, planes, and trains we take $\rho_{air} = 1.3 \times 10^{-3}$ g/cm³, $\eta_{air} = 1.8 \times 10^{-4}$ Poise, $v \cong 10 - 100$ m/sec $= 10^3 - 10^4$ cm/sec, and $d \cong 0.5$ m $= 50$ cm and then find $N_R \sim 1,000,000$, which is clearly in the inertial force domain where streamlining is important. On the other hand, for flying insects we find $N_R \sim 10$ where the weight added by streamlining would probably do more harm than good. Mother Nature has streamlined sparrows but not mosquitos, and we can understand why. We note that large insects (beetles and locusts) are streamlined.

The flow of fluids through tubes is a special case of fluid friction. We have seen that the influence of the stationary walls of the tube extends all the way to the center of the tube because the velocity distribution is parabolic everywhere and uniform nowhere. We might ask when fluid flow through a tube ceases to be laminar and becomes turbulent. The surprising answer is that the Reynolds number must reach the range of 10^3 before this happens. Of course, once the flow does become turbulent, its characteristic changes dramatically.

When fluid is forced through a tube by a pressure difference, the rate of flow is given by Poiseuille's law (Eq. 3.24) as long as the flow is laminar. Figure 3.20 shows that the flow rate is proportional to the pressure drop until

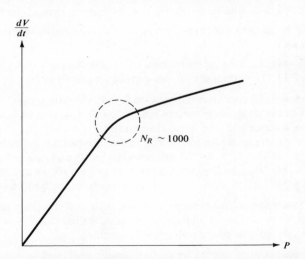

Figure 3.20 Plot of fluid flow in a tube versus applied pressure difference. The graph is linear with pressure in accordance with Eq. 3.24 for flow rates small enough to give small values of N_R. At larger flow rates the turbulence offers more resistance to flow than viscous friction and the graph is no longer linear.

N_R exceeds about 1000, and then the slope of the curve changes. It takes a much larger increase in pressure to double the flow when it is turbulent than when it is laminar. The work done by the pumping force is dissipated by turbulent flow in the form of excess heat, sound, and vibration.

In the human body, most fluid flow is laminar. Blood flow in the aorta at the peak speed is just barely laminar and in many cases becomes turbulent. Urine flow in the urethra is very often turbulent. There is a new diagnostic technique called *urodynamics* that studies the urine stream to learn about turbulence and obstructions causing it in the urinary system. Apart from these exceptions, nature has designed our fluid transport systems so that we dissipate a minimum of energy in overcoming turbulence.

Questions and Problems for Chapter 3

1. Suppose a cube of mysterious metal is suspended by a string attached to a scale that reads 100 N. The cube is now lowered into a tub of water and the scale reads 90 N. Why? How big is the cube? What is the density of the metal?

2. What is the difference in the fluid pressure between your head and your toes when you are standing? When you are sitting? What about for a giraffe?

3. The systolic pressure from your heartbeat is 120 mm Hg. What does that mean? What is the pressure in your head? In your feet?

4. Why are dams built with more thickness at the bottom than at the top?

5. Floating icebergs show only one-tenth of their volume above water. What is the density of ice?

6. How much weight can be supported on a cubic cork float of volume V? (Density of cork is 0.15.) How does the answer depend on the shape?

7. "High Density" Harvey likes his new backyard swimming pool, but he can neither swim nor float, so he uses an inflatable raft. One day while on the raft, he loses his balance and falls in (to the bottom).
   (a) Does the water level rise, fall, or stay the same when this happens? Explain.
   (b) Harvey's neighbor, Nick O'Tyme, hears the splash and runs to help. If Nick can lift only 20 kg on dry land, and Harvey weights 100 kg and has a density of 1.1 g/cm³, will Nick be able to lift Harvey from the bottom of the pool?

8. We have assumed that as the arteries stretch, their wall thickness remains constant. This is not so. Derive an expression for P_b for the case where t decreases as $1/r$; i.e., the volume of the wall remains constant.

9. Calculate the pressure exerted by surface tension on blood flowing through a capillary of diameter 0.001 cm. The surface tension of water is about 50 dynes/cm and blood is about the same. Make reasonable assumptions about any other numbers you need and give your answer in mm of Hg. (1 atm = 10^6 dyne/cm².) What effect does this have on circulation?

10. Two vertical plates are at an angle θ to each other as shown in the sketch. They touch at one edge, much like the pages of an open book standing vertically. They

rest in a pan of water and capillary action causes the water to rise between them. The height of the water is not the same everywhere because the plates are not everywhere equidistant. Derive an expression for the height of the water as a function of the distance from the edge where the plates touch (x in the sketch).

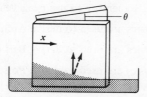

11. Why have we assumed in the text that the pressure required to inflate the alveoli (and therefore to breathe) is just equal to that produced by surface tension? Why shouldn't the pressure be slightly greater than this?

12. A fireman has a water supply at pressure P and would like to get it as high as possible. Should he stand on the ground with his hose and squirt upward or carry the end of the hose as high as the pressure P (at ground level) will allow? Neglect viscosity and friction.

13. Use Bernoulli's equation to find the force required to walk or ride a bicycle against the wind. Assume the wind speed is 10 mph and the density of air is $\frac{1}{750}$ g/cc (10 mph = 400 cm/sec). At what speed does the power requirement become as large as the 150 watts that a trained athlete can produce?

14. Consider the container in Figure 3.17. For every hole placed above $h = d/2$ there is a location below $h = d/2$ where another hole would result in the fluid being squirted the same distance x. Find a relation between these holes' heights.

15. A siphon is a device to transfer liquid to a lower place over a barrier (see sketch). How does it work? Explain what limits the flow rate.

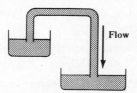

16. Why don't the smallest bronchi constitute a large fraction of the flow resistance? (Bronchi may be as small as 0.4 mm in diameter.) Knowing that a man's BMR O_2 requirement is 16 ℓ/hour, calculate the Reynolds number for air in the trachea, assuming that air is incompressible. Is this flow turbulent? Is the flow turbulent if the air flow is 10 times as great? Clearly state all assumptions you must make.

17. Derive an expression for the diameter of a falling stream of water as a function of height. What does this predict for very large heights? What do you expect will really happen? Why?

18. The inside diameter of a hypodermic needle is 0.01 cm and the viscosity of a saline solution is 0.01 Poise (cgs unit). If an intravenous infusion of 3 ℓ/day is to be administered by a bottle mounted $\frac{1}{2}$ m above a patient's bed, how long should the needle be? What limits the flow of an intravenous feeding system?

19. The heart pumps 70 cc of blood 72 times per/minute into a 1-in. diameter aorta. Find the Reynolds number. Is the flow turbulent? What are heart sounds and heart murmurs from? (Remember to use peak velocity, not average velocity.)

20. What principles of physics are used to derive Archimedes' principle?

21. What principles of physics are used to derive Bernoulli's law?

22. What principles of physics are used to derive Poiseuille's law?

Bibliography

1. G. AIELLO, P. LATRANCE, R. RITTER, and J. TREFIL, *Physics Today*, **27**, 23 (September 1974). (A description of the urinary drop spectrometer used for urodynamics.)

2. MARY ELLEN AVERY et. al., The Lung of the Newborn Infant, *Scientific American*, April 1973.

3. C. F. DEWEY and M. Y. JAFFRIN, Lecture Notes for Course in Biomedical Fluid Mechanics at M.I.T., Cambridge, Mass., 1971.

4. H. GOLDSMITH and R. SKALAK, "Hemodynamics," *Annual Review of Fluid Mechanics*, **7**, 213 (1975).

5. Russell HOBBIE, *Intermediate Physics for Biology and Medicine*, John Wiley & Sons, New York, 1978. (Contains a good discussion of fluid flow.)

6. LEONARD LEYTON, *Fluid Behavior in Biological Systems*, Clarendon Press, Oxford, England, 1975. (Detailed discussion of a number of topics in fluid dynamics applied to living beings.)

7. R. RESNICK and D. HALLIDAY, *Physics I* (John Wiley and Sons, Inc., New York, 1966). (A standard introductory text.)

8. I. RICHARDSON and E. B. NEERGAARD, *Physics for Biology and Medicine* (John Wiley and Sons, Inc., New York, 1972).

9. T. RUCH and J. FULTON, *Medical Physiology and Biophysics*, 18th ed. (W. B. Saunders Co., Philadelphia, Pa., 1960). (One of the all-time standards in physiology texts.)

10. F. SEARS and M. ZEMANSKY, *University Physics*, 4th ed. (Addison-Wesley Publishing Co., Inc., Reading, Mass., 1970). (A standard introductory text.)

11. E. SELKURT, *Physiology*, 3rd ed. (Little, Brown and Co., Boston, Mass., 1971). (A standard text.)

12. A. SHAPIRO, *Shape and Flow* (Doubleday and Co., Inc., New York, 1961). (This book is a superb guide to problems in fluid friction. Its clearly written exposition is indispensible for an understanding of drag forces.)

4

BLOOD CIRCULATION

Part 1: Introduction to the Circulatory System

Blood circulation has always been one of the most carefully studied aspects of human physiology. It has attracted attention because it is relatively easy to obtain a basic understanding of its function and because of its obvious importance to survival. The heart pumps blood throughout the network of blood vessels called the *circulatory system,* and this singular, simple function makes the heart more amenable to understanding than a chemically complex or multifunctioning organ such as the liver. Even our language reflects the importance of the heart: e.g., "Let's get to the heart of the problem."

The heart is a four-chambered pump (two ventricles and two atria) whose functions are schematically illustrated in Figure 4.1. Blood enters the left ventricle (LV) from the left atrium (LA) and it is pumped from there into the aorta, which divides the flow into several major branches. These branches serve the head and arms, liver and spleen, kidneys, stomach and intestines, and the muscles of the legs and trunk. The blood passes through the capillaries located in each of these major areas into the veins and is eventually collected in the right atrium (RA). From there it flows to the right ventricle (RV) where it is pumped through the pulmonary arteries to the lungs. After oxygenation in the lung capillaries of the alveoli, it returns through the pulmonary veins to the left atrium. Thus the cycle is complete.

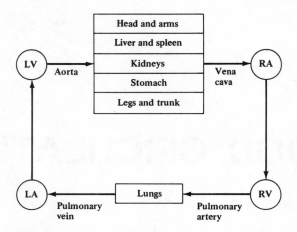

Figure 4.1 Highly schematic diagram of circulatory system showing function of each of the major chambers of the heart.

Of course, the heart's four chambers are not located in four different parts of the body as the Figure 4.1 implies. The right and left chambers are separated from one another only by a wall called the septum, as shown in Figure 4.2. Blood from the atria enters the ventricles through valves as shown and flows out to the great arteries through another set of valves. These arteries leave the heart near the atrial septum as shown in Figure 4.2.

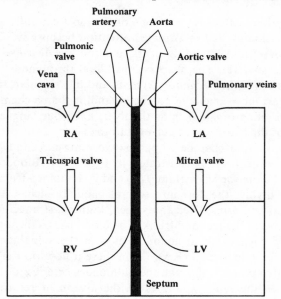

Figure 4.2 Highly schematic diagram of the heart showing blood flow between the major chambers.

The arrangement of the muscular walls of the heart is particularly well suited to a higher pressure in the left ventricle than in the right. This is necessary because the left ventricle pumps blood throughout the body whereas the right one pumps blood only to the lungs. Clearly a plane septum can not support any ventricular pressure difference unless it were rigid, which is not the case. In fact it is curved out into the right ventricle in order to maintain the highest possible concave curvature in the left ventricle. We have already seen in Chapter 3 that, for a fixed tension, a higher curvature (smaller radius) boundary produces a larger pressure. Figure 4.3 shows the shape of the walls of the ventricles as if the heart has been cut in a horizontal plane just below the atria. The walls of the left ventricle are also thicker and stronger than those of the right.

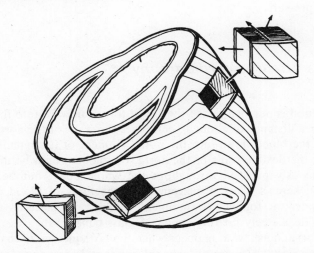

Figure 4.3 Sketch showing ventricular musculature. Note the direction of the muscle fibers and the shape of the two ventricles. (From Selkurt, *Physiology*, 3rd ed. © 1971 Little, Brown and Co. Reprinted by permission.)

The heart and the major vessels that transport blood form a rather complicated three-dimensional structure that is not easy to draw. The pulmonary artery leaves the right ventricle, passes up between the atria, crosses in front of the aorta, and curves to the left. Further along, the right pulmonary artery splits off and curves back behind the aorta to the right lung. The aorta comes up out of the left ventricle also between the atria, arches up over the right pulmonary artery, sends out three branches to the head and arms, and then goes down behind the heart to the trunk and legs. Since the aorta and the pulmonary artery are each 2- or 3-cm diameter vessels, and since the heart is only about as large as a closed fist, the interweaving of these arteries with the pulmonary veins, vena cavae, and cardiac blood supply vessels makes a rather complicated structure. The reader is referred to a good physiology text or encyclopedia for drawings.

All cyclic pumps require valves. The valves that keep the blood from flowing back into the ventricles after the ventricular contraction are called the *aortic* and *pulmonary* semilunar valves. The valves that keep the blood from flowing back into the atria during ventricular contraction are called *atrioventricular* (AV) valves, the one on the right being tricuspid and the one on the left being bicuspid (and sometimes called mitral). The rather soft cusps of the AV valves are closed by blood pressure during ventricular contraction (systole) but are supported by thin chordae-tendonae against the full force of the pressure. These chords, fastened to the ventricular walls, are of the right length to allow the valves to close completely but not invert under systolic pressure.

Part 2: Ballistocardiography

When blood is pumped out of the LV, it travels upward into the arch of the aorta for a distance of about 10 cm. Since the walls of the aorta are elastic, most of the blood flows into the initial segment of it stretching the walls. Without the flexible walls of the aorta, the blood would be uniformly and simultaneously pumped throughout the whole body, but because of the elasticity there is a net displacement of the mass of blood in the body.

Consider the following experiment. Suppose we place an individual of mass M on a freely floating couch of mass considerably less than M (see Figure 4.4) and ask what happens when the LV squeezes 70 cc of blood into the aorta. Since there are no external horizontal forces on the person-plus-couch combination, its center of mass must remain fixed. The coordinate of the center of mass x_{cm} is given by

$$x_{cm} = \frac{mx + MX}{m + M} \tag{4.1}$$

where small letters refer to the blood and capitals refer to the rest of the body and couch. The blood will move by Δx when the ventricles contract, and we require that the center of mass not move. We find

$$m\Delta x + M\Delta X = 0 \tag{4.2}$$

which says that the body and couch will move by an amount

$$\Delta X = \frac{-m}{M}\Delta x = -\frac{1}{1000}10 \text{ cm} = -0.1 \text{ mm} \tag{4.3}$$

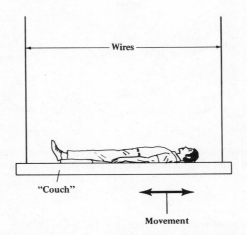

Figure 4.4 Subject on a home-built ballistocardiograph.

for $m = 70$ g, $M = 70$ kg. Since the ventricular systole lasts for about $\frac{1}{3}$ sec, the velocity is

$$\frac{\Delta X}{\Delta t} = v = 0.5 \text{ mm/sec} \qquad (4.4)$$

and the acceleration is

$$\frac{\Delta v}{\Delta t} = 2.5 \text{ mm/sec}^2 = \frac{-g}{4000}. \qquad (4.5)$$

Of course, all of these quantities average to zero over a full cardiac cycle.

It is relatively easy to measure displacements and velocities of the magnitudes indicated here, and a properly instrumented couch system is called a *ballistocardiograph*. Ballistocardiography is used by diagnosticians to measure cardiac output and other quantities associated with blood flow. In order to make the ballistocardiograph as sensitive as possible to the rather small movements and accelerations, the couch should be of the smallest possible mass. In order to keep the couch from moving too far in the event the patient moves (for example, coughs), it must be restrained by a restoring force or suspended as a pendulum. In either case, the system will have a resonance frequency and that frequency must be chosen to be far from the frequencies associated with the ballistocardiogram.

Clinical ballistocardiography is not very widespread but can be used to provide certain diagnostic information. Figure 4.5 shows a "normal" and pathologic ballistocardiogram. Such traces are of little general use, however, because small differences in the apparatus can make rather different looking traces. Until the hardware is standardized, ballistocardiography will continue to be a rare and esoteric diagnostic technique.

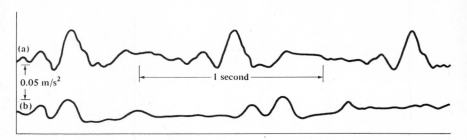

Figure 4.5 Ballisotocardiograms for (a) a healthy young subject and (b) a subject recovering from a heart attack. Note that the amplitude of (a) is larger than that of (b).

Part 3 :   Mechanical Events of the Cardiac Cycle—Heart Sounds

We now discuss the mechanical events of the normal cardiac cycle. As the blood pours into the right atrium from the vena cavae and into the left atrium from the pulmonary veins, the atrial walls stretch, the pressure builds up, and the AV valves open. A considerable amount of blood enters the collapsed (from previous cycle) ventricles and expands them to their relaxed size. The atrial systole that comes along soon does not contribute more than 20% addition to the blood already in the ventricles. After atrial systole, however, there is enough ventricular pressure rise from the start of ventricular systole to close the AV valves.

When the ventricles contract, their muscular tension rises to its maximum in a very short period of time ($\frac{1}{20}$ sec). This is too short a time for much blood to flow out of them and the process can best be labeled as an iso-volumetric pressure increase. The apex of the previously elongated heart moves up bringing the heart into a more spherical shape. During the next fifth of a second, the blood flows out of the pressurized ventricles while the muscular tension remains fairly high (see Figure 4.6). The rate of flow rises rapidly from zero at the start of the stroke to some maximum value and thereafter decreases steadily because of the back pressure from the blood in the distended aorta. After the ventricular systole is complete, the aortic valve closes and the stretched aortic walls force the blood through the rest of the body.

The pulsatile nature of the heart characterizes circulation throughout the arterial system. It is only through the elasticity of the arterial walls and the flow resistance of the capillaries that the pulsations are smoothed and the blood flow is steady. This smoothing is analogous to the filtering of an electrical *RC* circuit. Note that the presence of capacitance, namely elasticity or

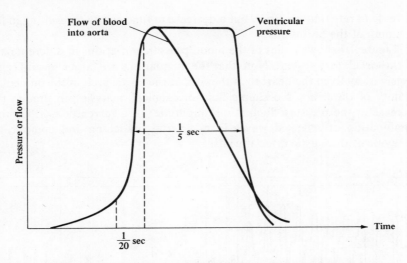

Figure 4.6 The time course of ventricular pressure and ventricular flow starting at the onset of ventricular systole.

compliance of the arterial walls, must be accompanied by a load resistance, namely viscous friction in the capillary bed.

We estimate the value of these parameters as follows. We know that the venous pressure is near zero and the arterial pressure typically varies from 120 mm at systole to 80 mm at diastole (ventricular relaxation). We say that the blood pressure is 120/80. The average pressure difference is 100 mm and it produces a blood flow of about 70 cm³/sec. From this we calculate a total peripheral resistance (TPR) of pressure/flow = $100/70 \cong 1.5$ mm Hg-sec/ cm³. To find the capacitance we assume that the 40-mm pressure rise of ventricular systole can deposit the entire stroke volume of 70 cc into the elastic aorta. The capacitance is then volume/pressure = 70/40 = 1.75. Note that the pressure difference used to estimate the capacitance is not the same as the pressure difference used to estimate TPR. The time constant is the product $(1.75)(1.5) \cong 2.7$ sec, which is about the time for three complete heart cycles. Since an RC filter reduces the amplitude of an ac signal by a factor $(1 + 2\pi f RC)$, we see that the aortic elasticity and TPR combine to reduce pressure fluctuations in the veins by more than a factor of 20.

Heavy exercise results in an increased demand for tissue oxygenation, which is met by increased cardiac output. The blood vessels dilate resulting in a reduced TPR so that the blood flow rate increases. The reduced TPR also reduces the time constant of the RC filter so that the pulsatile flow is not as well damped, and that is why we can feel the pounding of our heart so much more strongly after a bout of heavy exertion. It is not simply an increase in rate that we sense, but a decrease in time constant resulting in less damping of the pulsations.

Figure 4.7 shows a plot of the blood pressure variations in different parts of the circulatory system. Note that the fluctuations are larger several centimeters away from the heart than they are at the aortic arch in the immediate vicinity of the heart. No simple fluid mechanical analysis can predict this increase in the pressure fluctuation amplitude; it is currently ascribed to a combination of reflected waves from arterial branching and complicated behavior of the elastic arterial walls.

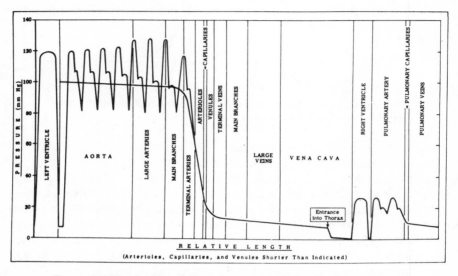

Figure 4.7 Blood pressure in various parts of the circulatory system. Although the horizontal axis represents distance, the variations of pressure shown in the arterial system represent temporal rather than spatial fluctuations. The pulse is damped primarily in the arterioles and capillaries. (From Selkurt, *Physiology*, 3rd ed. © 1971 Little, Brown and Co. Reprinted by permission.)

The heart sounds arise from the opening and closing of the various valves and from the rapid movement of blood. The first and loudest sound comes from the closing of the AV valves after atrial systole. The mitral valve closes about 10 msec before the tricuspid, but the closing event itself takes about 30 msec. If the right and left ventricles begin systole at different times for some reason (nerve damage, muscle defect), this interval is changed and can often be detected using a stethoscope as a splitting of the first sound. The movement of blood into the arteries also contributes to the first sound, but not until some time has elapsed after the valve closings. If there is a circulatory problem that produces considerable turbulence associated with this flow, the sound is called a heart murmur associated with first sound. The first sound is therefore a complex one arising from a variety of sources.

As the pressure builds in the left ventricle during the iso-volumetric contraction phase, it eventually reaches the $\cong 80$ mm Hg of the aorta (still distended from the previous cycle) and the aortic valve opens allowing flow out of the ventricle. The ventricle collapses rapidly as it ejects blood into the expanding aorta and eventually the pressure in the aorta becomes as large as that in the ventricle. The aortic valve then closes producing the second heart sound. A similar process associated with the pulmonic valve and right ventricle produces a related sound. The combined sound from both semilunar valves is simpler than the first sound because it is not accompanied by the initiation of any blood flow. Nevertheless, the separation of sounds from the aortic and pulmonic valves varies with respiration, making the second sound somewhat more difficult to interpret than the first. A variety of pathological situations including ventricular septal defect, mitral valve regurgitation, aortic stenosis, and cardiac failure can be diagnosed from the respiratory variations of the splitting of the second heart sounds.

The third sound (generally inaudible) is derived from the flow of blood from the atria to the ventricles. Since this does not occur under the pressure of any contraction nor is there any valve closing associated with it, it is much weaker.

There are many audible sounds that derive from blood flow through and around obstacles in the vessels rather than from the beating of the heart. When vascular constrictions or obstructions cause the velocity of blood flow to increase substantially, the Reynolds number can easily exceed 1000 and the flow becomes turbulent. Noise resulting from this turbulence can be heard as a rushing sound at each systole. The nature of these sounds can lead to a variety of diagnoses.

The power spectrum of heart sound is generally confined to frequencies below the audible range, for the most part, below about 30 Hz. Figure 4.8 shows the approximate power distribution of heart sounds along with a typical human auditory threshold curve. The very small region of overlap shows why most of the information from heart sounds is not detectable by the human ear; for this reason, phonocardiography is still rather primitive. It is necessary to use special instruments to detect these low-frequency sounds. Since most acoustic research today is directed toward the use of ultrasound (higher frequencies), it seems that this area will continue to be neglected.

In addition to phonocardiography, there is a large number of other cardiodiagnostic techniques using sound waves. These methods use ultrasonic (high-frequency) waves that are sent into the body by small transducers placed on the skin. Studies of the echoes provide a great deal of information about the heart and will be discussed in Section 9.7.

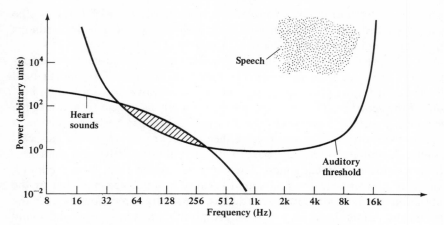

Figure 4.8 A spectral diagram of heart sounds. The area below the curve labeled heart sounds represents their domain, which is seen to be primarily at low frequencies. Only a small fraction of their power is above the threshold of hearing, which means special instruments are needed for their detection and study.

Part 4 :   Electrocardiography

The electrocardiogram (ECG or EKG) is a series of recordings of temporal voltage variations between various surface points on the body. These voltages are recorded with a moving strip of paper and a pen (strip-chart recorder) in such a way that the time dependence is displayed on the paper. These voltages originate from the electrical activity associated with the action of the heart muscles.

Figure 4.9 shows the temporal sequence of most of the events of the normal cardiac cycle, including heart sounds and valve openings and closings. At the top of the figure is a typical electrocardiograph (ECG) trace. The various features of the trace have historically been given the names of the letters P through T. The cycle is begun by an electrical pulse from a bundle of specialized muscle cells called the sinoauricular (SA) node, which is the pacemaker of the heart. The output of the SA node causes atrial depolarization (P wave) initiating atrial systole. The pulse is also transmitted to another ganglion of cells called the atrioventricular (AV) node where it is delayed somewhat but then retransmitted along the ventricular septum by a collection of fibers called the bundle of His. The wave of ventricular muscle depolarization (for a discussion of cell polarization, see Chapter 6) travels down the septum to the apex and then spreads out on the outside surface of the apical

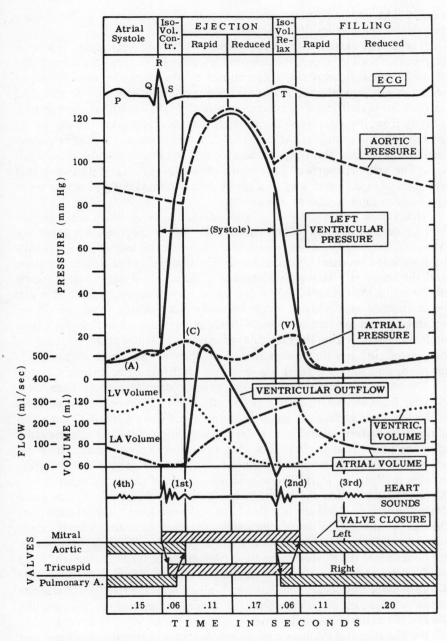

Figure 4.9 Time course of the cardiac cycle showing the ECG, valve openings and closings, and various pressures and flows. (From Selkurt, *Physiology*, 3rd ed. © 1971 Little, Brown and Co. Reprinted by permission.)

region. The depolarization causes a sequential contraction of the various ventricular muscles resulting in ventricular systole.

The electrical activity preceding the very strong ventricular systole is responsible for the large QRS component of the ECG. The relative size and shape of the various peaks is determined by the electrode locations as we shall see, but generally the R wave is the largest. In addition to the electrical activity from ventricular depolarization in the QRS complex, there is also activity from the atrial repolarization. It is masked, however, by the strong signal from the ventricles. After the QRS complex there is a relatively quiet period followed by the T wave that accompanies the ventricular repolarization. The time dependence of the events of repolarization and depolarization are somewhat different, and as a result their accompanying components of the ECG are not similar to one another.

Heart muscles are made of a large number of small fibers that are normally polarized by about 70 mV across their membranes (inside is negative). Muscular activity is started by depolarization of the individual fibers. We have already seen that the depolarization is not simultaneous for all the fibers and the result is a changing pattern of polarized and depolarized muscle fibers during the course of the cardiac cycle. Each polarized muscle fiber can be considered as having a small dipole moment arising from the charge separation, and the sum of all these dipole moments is a macroscopic dipole moment characterizing the particular phase of the cardiac cycle. As the cycle progresses, the position and strength of each of the dipole moments is changed, and thus the heart dipole vector is changed [see Figure 4.10(b)].

The usual course of the wave of heart muscle depolarization is as follows. A pulse from the SA node begins the atrial depolarization that progresses along the outer surfaces of the atria [1 and 2 in Figure 4.10(a)]. Conduction along these surfaces is so slow that the depolarization of the left atrium occurs after the pulse from the AV node that starts ventricular systole. There are few muscle fibers in the atrial septum because of the presence of the great arteries between the atria. The depolarization of the ventricles begins in the AV node deep within the heart. The wave of depolarization is conducted down the septum along the bundle of His and also along the front surface of the right ventricle. The wavefront reaches the apex somewhat after the right ventricle has been depolarized [3 and 4 in Figure 4.10(a)]. The apex depolarization results in the R wave of the ECG (the largest peak) and is then followed by the depolarization of the rest of the left ventricle [5, 6, and 7 of Figure 4.10(a)]. Since this occurs simultaneously along both the front and back side of the ventricle, the electrical effects are somewhat canceled and the resulting S wave is smaller than the R wave.

We visualize this event as a single dipole vector centered at the heart and rotating at an uneven rate through 360° with each cycle [Figure 4.10(b)]. Of course, the magnitude of the vector can change as well as its direction. An

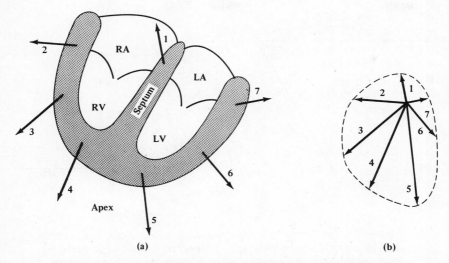

Figure 4.10 The electric dipole moment of the heart originating from different places at different times of the cycle is shown in (a). The various electric dipole vectors are drawn with a common origin in (b).

electric dipole located in the midst of a bulk conductor such as the human trunk establishes currents in the bulk. These currents also flow at the surface where the voltages they produce can be measured with electrodes. A record of the time dependence of these voltages constitutes the ECG.

Since ECG's are usually taken with three leads attached to the two arms and the left leg, we consider the voltage difference between different pairs of points chosen from among these three. Since there is no source of voltage at the ends of these limbs, it is clear that there is no surface (or bulk) current flowing and therefore the voltage must be the same at the wrist as it is at the shoulder. There are three such pairs of points, and Kirchhoff's laws require that the sum of the three voltage differences is zero. By measuring the voltages at the wrists and left leg, one measures the voltages at the shoulders and lower abdomen. These three points define the *Einthoven triangle*. There is a convention for labeling the signals taken between the points of the triangle:

$$\text{Lead I} = \text{left arm} - \text{right arm} = \text{LA} - \text{RA}, \qquad (4.6a)$$

$$\text{Lead II} = \text{left leg} - \text{right arm} = \text{LL} - \text{RA}, \qquad (4.6b)$$

$$\text{Lead III} = \text{left leg} - \text{left arm} = \text{LL} - \text{LA}. \qquad (4.6c)$$

It is clear from this convention that Kirchhoff's law requires $V_{\text{I}} + V_{\text{III}} = V_{\text{II}}$.

We imagine the heart's dipole vector to be projected against the sides of the Einthoven triangle as shown in Figure 4.11(a). The little spikes in Figure

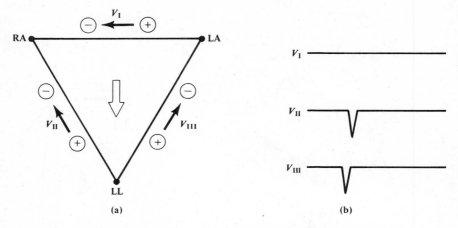

Figure 4.11 The Einthoven triangle with the polarities of the leads indicated by arrows that point from negative to positive is shown in (a). Part (b) shows the pulse expected at each lead for the dipole vector shown in heavy ink in (a). Because the vector is perpendicular to the axis of lead I, there is only a signal on II and III. Since the dipole points opposite to the polarity arrow, both spikes are negative.

4.11(b) are intended to indicate the voltages measured at a particular instant as a result of an instantaneous electric dipole moment represented by the heavy arrow in Figure 4.11(a). Observe how the condition imposed by Kirchhoff's law is met.

In the series of sketches in Figure 4.12(a), we see how various cardiac dipole vectors might project against the sides of the Einthoven triangle. In Figure 4.12(b) is a typical sequence of ECG recordings as they might normally appear when the recordings from each of the three pairs (loosely called "leads") are plotted on the same axis. Notice that the R wave is very large and positive on both I and II because of the sign convention in the definition of II. Notice that $V_I + V_{III} = V_{II}$ in both Figures 4.12(a) and 4.12(b).

Usually there are other recordings in addition to the limb records associated with the Einthoven triangle. For these, the three limb leads are combined so that their voltages are summed and the resultant is considered an electrocardiographic body ground. The voltage differences between this ground and various points along the chest are recorded and given arabic subscripts. The locations of the points are shown approximately in Figure 4.13(a). In Figure 4.13(b), the typical tracings obtained from these leads are shown. Notice the progressions of the various components of the signal.

The measurement of the various projections of the cardiac dipole moment in several body planes provides gross, averaged information about the cardiac vectors. The process is often called vector cardiography. A trained diagnostician who has studied a very large number of normal and abnormal

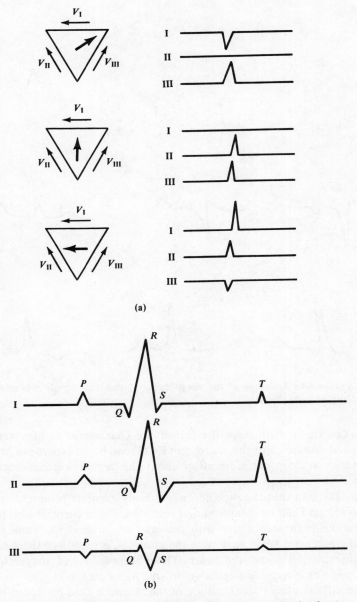

Figure 4.12 The Einthoven triangle with various cardiac dipoles shown at the left gives the signals shown at the right in (a). A typical ECG trace for each lead is shown in (b).

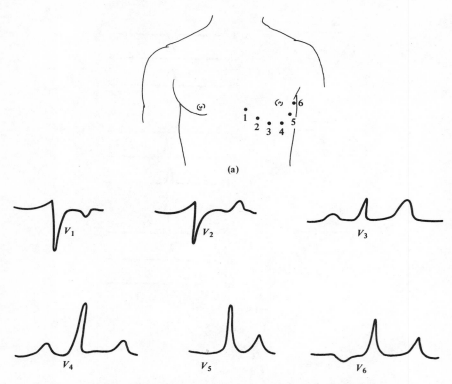

Figure 4.13 Location of the chest leads (a) and the signals typically derived from them (b).

ECG's can easily distinguish the identifying characteristics of a variety of defects and diseases. For this reason, an ECG read by an expert can provide a tremendous amount of information about the general cardiac status of a patient.

We have seen that the time-dependent body currents produced from the cardiac voltages can be measured and recorded. These currents also produce magnetic fields in accordance with the laws of classical electromagnetism. Since the magnetic fields vary with the currents, a record of them can also carry information about the heart. These records, called magnetocardiograms, are slowly gaining popularity in study and diagnosis.

We shall calculate the magnitude of the magnetic field produced by these currents. We approximate a region of the body just under the skin near the heart as a thin sheet of current. If we consider the magnetic field very close to this surace, we can use Ampere's law of electromagnetic theory to find the magnetic field $B = 2\pi \times 10^{-3}(I/d)$ gauss where I is the current (amperes) spread out over a sheet of width d (meters). The field direction is parallel to the skin but perpendicular to the current flow. In order to estimate I we use the fact that the body resistance over a centimeter size region is about 100 ohms and that the peak ECG voltage is about 10 mV. We find $I \cong 0.1$ mA. We choose d to be a strip about as wide as the heart ($\cong 0.1$ m) and find $B \cong 6 \times 10^{-6}$ gauss.

Fields as small as this are very difficult to detect with ordinary instruments in ordinary circumstances. The earth's field is typically 0.5 gauss and fluctuates at frequencies near those of interest to magnetocardiography by about 10^{-4} gauss. External disturbances (motors, instruments, etc.) can also produce fluctuating fields that, like the earth's field fluctuations, could swamp a magnetocardiograph. It is necessary to build specially shielded rooms to study the body's magnetism. Furthermore, once the effects of external magnetic noise have been shielded out, it is necessary to use very sensitive magnetic probes. The most sensitive devices we know are Superconducting QUantum Interference Devices (SQUID), which are quite adequate for the job. In Figure 4.14 a sample of magnetocardiograph signals are shown superimposed on an outline of the heart. The P, QRS, and T waves show up quite clearly,

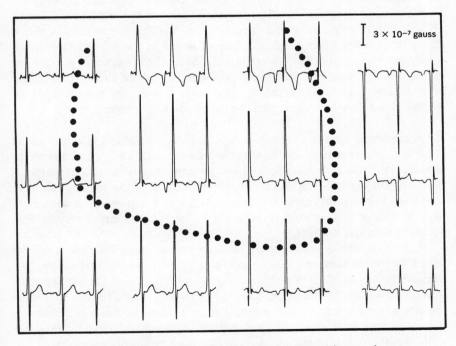

3×10^{-7} gauss

Figure 4.14 Magnetocardiogram signals from various points on the surface of the chest with a superposed outline of the heart. Notice the presence of the same waves that are found in the ECG. (From Cohen, *Physics Today*, **28**, 34, August, 1975. Reprinted by permission.)

although with different amplitudes and phases than we would expect since they are taken at different points. We see, for example, that the QRS complex is very large near the apex and is dramatically smaller near the atrial regions and at some distance below the apex. The dashed line in Figure 4.14 divides the regions of different phases for the QRS complex. Note that a similar line could be drawn to divide the regions of opposite phase for the P wave, but it would be a different line because this signal arises from a different part of the heart.

Part 5: Hemodynamics

The flow of blood in the human circulatory system exhibits a variety of characteristics that differ from those we have already discussed. It is important to keep in mind the parameters of blood flow in different parts of the body. For example, the peak flow velocity in the aorta is found by dividing the cardiac output (about 5 liters/minute) by the cross-sectional area (about 5 cm²) remembering that the flow is pulsatile and that the aorta is filled in about one-fifth of the time of a cardiac cycle. We find the velocity is about 80 cm/sec. On the other hand, the flow of blood in the capillaries is farily steady but still amounts to 5 liters/minute. Since the diameter of a capillary is only about 8 microns, but there are about 5×10^9 of them, the average flow velocity is about 0.03 cm/sec, 2500 times slower. We now proceed to discuss blood flow in various parts of the circulatory system.

A. Ventricular systole produces a rapid injection of blood into the aorta. Since the resistance to blood flow in the lesser vessels is high and the walls of the aorta are elastic, the blood does not immediately flow to all parts of the body but expands the aorta. The kinetic energy of the moving blood is converted to potential energy of the aortic wall, and that energy is converted back to kinetic as it is used to push the blood into the circulatory system during ventricular diastole.

If the flow into the aorta could be described as Poiseuille flow, it would have a Reynolds number of about 2000 and we might expect turbulence, but because the length of aorta that accepts pulse of blood from the left ventricle is only a few times its diameter, there is not enough length for the parabolic velocity distribution to become established, and the blood flows as a bolus or slug, with very little shear. Of course there is tremendous shear along the walls, but there is so little volume there that the amount of energy dissipated by turbulence is relatively small.

The cardiac output is limited to some extent by the back pressure exerted on the heart by the blood in the distended aorta. In fact, ventricular systole terminates when this pressure rises to equal the ventricular systolic pressure. When the body undergoes vigorous exercise and demands more blood, the peripheral vessels dilate allowing the blood to flow more easily to them. This dilation has the secondary effect of allowing more blood to be present in the periphery and therefore reducing the amount in the aorta at the onset of ventricular systole. The result is that the ventricle can push more blood into the aorta with each contraction because it takes a larger additional volume to reach the pressure needed to terminate systole.

A patient who has inadequate cardiac output because of a heart attack (loss of function of some ventricular muscle as a result of a partial blockage of the myocardial blood supply) is in danger of damage to the brain or other vital organs. It is possible to provide some assistance to the weakened left ventricle by artificial methods. One such technique, called a "counter-pulsating intra-aortic balloon" utilizes a rubber balloon inserted into the aorta through a catheter from the leg (femoral artery). During ventricular diastole the balloon is inflated forcing blood out of the aorta. The balloon is deflated during systole allowing more space in the aorta for blood storage so that the weakened heart can provide adequate blood without working quite so hard: The inflation-deflation mechanism is triggered by the ECG in order to achieve proper timing.

There are variations to this technique that are equally successful. It is possible to draw blood from the aorta into a reservoir during systole and return it during diastole for distribution to the body. It is also possible to squeeze the legs of a patient during diastole so that the aortic pressure will rise and the aorta will be already stretched before the systole. Needless to say, these methods of artificial assistance are only for emergencies and can only be used for a few hours immediately following a heart attack. Because the heart can regain some of its function in a relatively short time, and because the body can adapt to a reduced blood supply (rest is obviously needed), these artificial methods have been very effective in saving the lives of heart attack victims during the first few critical hours.

B. Flow in the arteries is much less complicated than in the aorta, but there are still a number of features to be discussed. Because the arterial walls are somewhat less elastic than the aortic walls, the pressure amplitude of the pulse is somewhat larger than it is in the aorta (see Figure 4.7). These pulse waves are complicated by reflections from arterial branches that can super-pose with the forward pulse wave and establish standing pressure waves. The result can be a rather complicated pressure distribution.

Let us consider what happens at an arterial branch (Figure 4.15). For simplicity we shall consider the case of a vessel of radius r that divides into two smaller ones of equal radius r'. In order for the smaller vessels to have the same viscous resistance/unit length, the radii must differ by a factor of

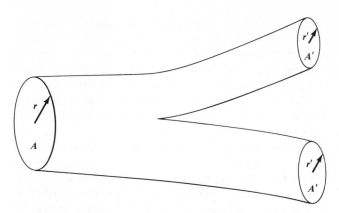

Figure 4.15 A bifurcation in a blood vessel. The two branches may or may not have the same cross-sectional area as the original blood vessel.

$\sqrt[4]{2} \cong 1.2$ (see Eq. 3.24). This means that the areas of the smaller vessels are each about 1.4 times smaller than that of the original, but since there are two of them, the total cross-sectional area is increased by $2/1.4 \cong 1.4$. This means that the flow velocity will be smaller. In addition, certain simplifying assumptions enable a calculation that shows that the Reynolds number is reduced by a factor $n^{3/4}$ where n is the number of vessels the original splits into. For the symmetrical bifurcation discussed above, N_R is reduced by a factor of $2^{3/4} = 1.68$.

Of course, it may be that the number and size of the smaller vessels is not chosen to offer the same resistance to Poiseuille flow. For example, more than two-thirds of the resistance to blood flow occurs in the arterioles and capillaries: there is only a small fraction of the total resistance in the major vessels. We must be very careful in considering the effect of the changing Reynolds number on the presence or absence of turbulence. Although N_R can be as large as 1000 and still allow laminar flow in straight, smooth tubes, the disturbance set up in a flowing fluid by a branch may very well induce turbulence at lower Reynolds numbers.

Turbulence can be artificially induced in arterial flow by external pressure, and such turbulence provides the most common method of measuring blood pressure. An inflatable cuff is wrapped around the arm and pressurized until it stops the arterial flow. A stethoscope is held over an artery down-stream of the cuff and the pressure is slowly reduced in the cuff until a little bit of blood can just barely spurt through the constricted artery at the peak of systolic pressure. The turbulent sound is detected and the pressure in the cuff is called the *systolic* pressure. The pressure in the cuff is lowered further until each pulse no longer produces a turbulent sound. This occurs when the cuff pressure is so low that it does not compress the artery even against the lower diastolic pressure. A blood pressure reading of "120 over 80" or 120/80 means systolic pressure is 120 mm Hg and diastolic pressure is 80 mm Hg.

The origin of the sounds, called *Korotkoff sounds,* is a matter of current debate but they are approximately associated with the events described above. The detailed questions surrounding their causes cast some doubt on the accuracy of blood pressure measurements with conventional cuffs. People are always seeking new and better ways to measure blood pressure.

C. Almost one-third of the pressure drop in the circulatory system occurs in the arterioles. These tiny vessels, about 30 microns in diameter, are interwoven with the capillaries in the tissue of the body and serve to distribute the blood into large regions. A calculation of the expected pressure drop in the arterioles as a result of viscosity leads to a much larger drop than that which is measured.

In 1931 a paper was published by Fahraeus and Linquist in which they described some rather peculiar properties of blood flow in smaller vessels (arterioles). They found that there was an apparent decrease in the viscosity of blood when it flowed in arterioles and that the hematocrit of the blood in the arterioles was less than that of the blood in the rest of the system. (*Hematocrit* is the ratio of red blood cell (RBC) volume to total volume and is determined by centrifugation of a blood sample to precipitate the RBC.) It is very difficult to imagine what could happen to the blood in the arterioles to make the apparent viscosity less than what it is in larger vessels, and it is absurd to believe that there are fewer RBC in arteriolar blood.

The model proposed to describe the Fahraeus-Lindquist effect is the following. Suppose the RBC's were concentrated toward the central part of the vessel. Then there would be fewer of them near the walls and the blood flow would be lubricated a little by the boundary layer of liquid since the plasma without the RBC's has lower viscosity than the whole blood. Furthermore, since the fluid moves faster near the center of the vessel, the same number of RBC's can be carried by the flowing blood at a lower average density (hematocrit) than the static hematocrit.

The model can best be understood by considering an extreme situation. Suppose there is a blood vessel carrying pure plasma, but it has a channel left open in the center, and RBC's travel at very high speeds through this channel. If the speed is high enough, it's possible to have only one RBC at a time in a section of vessel, but the net flux of them would be high enough so that, if they are combined with the slowly flowing plasma, the hematocrit is that of whole blood. In the hypothetical blood vessel, however, there is only one RBC at a time traveling very fast and the hematocrit is very low.

With this explanation of the extreme case of the model, we now proceed with a quantitative study. We first define the hematocrit ratio $h = $ RBC volume/total volume. For a current-carrying blood vessel, the output blood has a hematocrit $h = $ volume/second of RBC's divided by fluid volume/second. We now consider the case where the hematocrit is not uniform throughout the volume but varies only with radius; that is, $h = h(r)$. Then the amount of RBC's in any cylindrical shell of volume dV is $h(r)\, dV$. (We recall that the laminar flow of fluid in a smooth straight pipe is described as a series of concentric cylindrical shells, each moving with different velocities.) The velocity distribution given by Eq. 3.22 is

$$v(r) = \frac{\Delta P}{4\eta L}(R^2 - r^2). \tag{4.7}$$

The volume flow of any cylindrical shell is

$$v(r)\, dA = 2\pi r v(r)\, dr \tag{4.8}$$

and the RBC flow of a cylindrical shell is

$$h(r)v(r)\, dA = 2\pi r h(r)v(r)\, dr. \tag{4.9}$$

We calculate the total flow and total RBC flow by integrating these expressions and divide them to find the hematocrit of the transported blood. This must be equal to the average hematocrit h. We have

$$h = \frac{\int_0^R h(r)v(r)\, dr}{\int_0^R v(r)\, dr} = \frac{\int_0^R h(r)(R^2 - r^2)r\, dr}{\int_0^R (R^2 - r^2)r\, dr} = \frac{\int_0^R h(r)(R^2 - r^2)r\, dr}{\frac{R^4}{4}}. \tag{4.10}$$

What is the form of $h(r)$? Let us suppose it is parabolic just as the velocity distribution and that it is zero at the walls and has some maximum value H at the center. Then $h(r) = H[1 - (r/R)^2]$ and we can evaluate the integral in the numerator of Eq. 4.10 for h. The result is $HR^4/6$. Then we find $H = 3h/2$ for the peak value. We can also find the average hematocrit in the vessel itself. The total RBC volume is just

$$\int_0^R h(r)\, dV = \int_0^R h(r)2\pi r\, dr L \tag{4.11}$$

and the total volume is $\pi R^2 L$ so the hematocrit ratio is

$$\frac{\int_0^R H\left[1 - \left(\frac{r}{R}\right)^2\right]2\pi r\, dr L}{\pi R^2 L} = \frac{H}{2} = \frac{3}{4}h. \tag{4.12}$$

The hematocrit of the flowing blood is lower than h!

Let us now consider the case where the hematocrit is an exponential function of r instead of parabolic. In this case it can never become zero at the wall. We choose the exponential decay length to be just equal to R and then $h(r) = H \exp(-r/R)$. The denominator of the expression for h in Eq. 4.10 remains $R^4/4$ but the numerator becomes

$$H \int_0^R r(R^2 - r^2)\, e^{-r/R}\, dr. \tag{4.13}$$

This can be evaluated using standard tables and the result is $(14/e - 5)HR^4 \simeq 0.150HR^4$. Then we find $h = (\frac{2}{3})H$ or $H = (\frac{5}{3})h$. At the wall, the hematocrit $h(R) = 0.589h$. We can also calculate the average hematocrit in the flowing blood just as we did before. Again the volume is $\pi R^2 L$ and the hematocrit ratio becomes

$$\int_0^R 2H\, e^{-r/R} \frac{r\, dr}{R^2} \qquad (4.14)$$

which is also evaluated with tables. The result is $2H(1 - 2/e) = 0.528H = 0.88h$. Again we see the hematocrit is less than h in the vessel, but the final hemocrit of the transported blood is the same as that of the still blood.

We might speculate that the reason for the red blood cells being concentrated in the center of the vessel is that the pressure is lower toward the center because the fluid is flowing faster. The radial pressure gradient would tend to push the RBC's toward the center. This line of thought is not correct, however, because it is based on the notion that the pressure is lower where the velocity is higher in accordance with Bernoulli's equation. But the equation can only be applied to two points on the same streamline, and the streamlines do not run from the center to edge of the blood vessels. The concentration of suspended particulate matter toward the center of flowing fluids has also been observed in suspensions of clay in water and suspensions of glass beads. The physical processes underlying the Fahraeus-Lindquist effect are very subtle and are not simply explained.

D. Blood flow in the capillaries (microcirculation) is a very complicated and difficult topic that is still under active investigation. Since red cells are larger than the diameter of the capillary tubes, they must bend and fold to pass through. If the flexibility of the red cell wall is reduced (sickle cell anemia), a variety of problems arise.

Although the red blood cells must pass through the capillaries in a single file, it is not clear what the flow characteristics of the plasma must be. The naive picture of a bolus of plasma flowing along between the cells is probably wrong. Whether any plasma passes between the cells is not known, but there must be some layer of fluid between the RBC's and the capillary walls for lubrication. How this layer moves about and is maintained is a matter of some debate. However, the peculiar shape of the red blood cell (disc with a depressed center) probably plays an important role in microcirculation.

E. Blood flow in the veins is fairly well described by our earlier studies of fluid flow. There is almost no pulse, the flow is steady and smooth, the walls are flexible but generally maintain constant vessel radius because of the lack of pulse, and the flow is generally slow enough so that there is no turbulence. For example, the largest Reynolds number in the venous system occurs in the vena cava, which is 20% larger than the aorta and carries a steady (nonpulsatile) flow. The result is an N_R that is lower by a factor of 6 than the aortic value. The veins occupy a considerable volume and contain most of the body's supply of blood.

Questions and Problems for Chapter 4

1. Based on the calculation done for ballistocardiography, can you expect to see your weight fluctuate as you stand on a scale? Justify your answer.

2. Suppose you wished to build the ballistocardiograph illustrated in Figure 4.4. What should you consider in choosing the length of the wires? How long should they be?

3. Calculate the kinetic energy imparted to the blood by the left ventricle when the body is at rest. Compare this with the work done against the aortic pressure.

4. During vigorous exercise, the ventricular stroke volume and the heart rate both double resulting in a fourfold increase of cardiac output. Repeat Exercise 3 for this case and discuss the results.

5. The human body contains about 6 liters of blood. What is the time for blood to make a complete circuit through the body at rest and during exercise?

6. Calculate N_R for flow in various parts of the circulatory system. Show that, for constant viscous resistance, $N_R \propto n^{-3/4}$ where n is the number of smaller vessels derived from one large one.

7. Explain why flow in the vena cava isn't turbulent.

8. Show that an RC filter reduces the amplitude of an ac signal by the factor $(1 + 2\pi fRC)$.

9. Suppose the aorta were empty at the onset of systole. What length would a 70-cc volume fill? Suppose it were filled to this length. By how much would its diameter increase to accommodate the additional 70 cc in the same length?

10. Suppose the radial distribution in RBC's in a small vessel was linear instead of parabolic or exponential. Calculate H, the maximum value of h in the center, and the average hematocrit of the flowing blood.

Bibliography

1. A. C. BURTON, *Physiology and Biophysics of Circulation* (Yearbook Press, Chicago, Ill., 1972).

2. D. COHEN, *Physics Today*, **28**, 34 (August 1975). (Describes various biomagnetic phenomena, including magnetocardiography.)

3. P. DAVIDOVITS, *Physics in Biology and Medicine* (Prentice-Hall, Inc., Englewood Cliffs, N.J., 1975).

4. FARHAEUS and LINDQUIST, *Am. J. Physiology*, **96**, 562 (1931).

5. H. GOLDSMITH and R. SKALACK, *Annual Review of Fluid Mechanics*. **7**, 213 (1975). (A discussion of several topics in hemodynamics.)

6. Hewlett Packard Corp., Palo Alto, Calif., Application note AN 710, Pressure Measurement.

7. ———, Palo Alto, Calif., Application note AN 711, Electrocardiograph Measurement.

8. ———, Palo Alto, Calif., Application note AN 712, Heart Sound Measurement.

9. RUSSEL HOBBIE, *Intermediate Physics for Medicine and Biology*, John Wiley & Sons, N.Y., 1978. (Contains an excellent discussion of electrocardiography.)

10. H. METCALF, *The Physics Teacher*, **10**, 98 (1972) (Tells how to build an ECG amplifier for less than $10.)

11. V. S. MURTHY et. al., *J. Biomechanics* **4**, 351 (1971). (The intra-aortic Balloon for Left Heart Assistance.)

12. D. R. RICHARDSON, *Basic Circulatory Physiology* (Little, Brown and Co., Boston, Mass., 1976.)

13. S. RIEDMAN, *Heart* (Western Publishing, Racine, Wisconsin, 1974). (An excellent introduction to cardiac anatomy and physiology.)

14. T. RUCH and J. F. FULTON, *Medical Physiology and Biophysics* (W. B. Saunders Co., Philadelphia, Pa., and London, England, 1960). (One of the all-time standard physiology texts.)

15. A. VANDER, J. SHERMAN, and D. LUCIANO, *Human Physiology: The Mechanisms of Body Function* (McGraw-Hill Book Company, New York, 1970). (A standard text.)

5

FEEDBACK
AND CONTROL

Part 1 : Introduction

All animals and most machines can function only when the values of certain parameters are held within a certain range. Humans must have their body temperatures regulated to within a few degrees, computers must be supplied with specific voltages, and automobile carburetors must have a certain level of gasoline in the float bowl. Under normal operating conditions these parameters would often drift out of the proper range if there were no means for controlling them. The regulation and control of various quantities is a vital function of living systems as well as of man-made machines, and its study forms an entire branch of mathematics and engineering. We shall discuss a variety of control systems with particular emphasis on those that regulate the functions of the human body.

Sometimes parameters are adjusted, or set, by means of controls that can only be manually changed. For example, the temperature of the water in an ordinary shower is determined by the positions of the hot and cold faucets. If the hot water pressure drops resulting in a drop of hot water flow and thus of the water temperature in the shower, the problem can only be corrected by manually changing the faucets. Such a system is an example of an *open-loop* control. Maintaining a constant temperature depends on the manual intervention of some person who determines that the water is too hot or too cold and then compensates for the error. The action of the person constitutes the closing of the control loop—if the person is reasonably sensitive and quick, the water temperature can be regulated quite well.

If there was an electric thermometer in the shower that could sense a temperature change and adjust the faucets to compensate, we would say that the control system was *closed-loop* or automatically controlled. Of course, any control system must have a closed loop, but we distinguish between those having a mechanical or electrical device and those requiring intelligent intervention. An example of closed-loop control can be seen in the system to regulate the height of water in a vessel illustrated in Figure 5.1. There is a

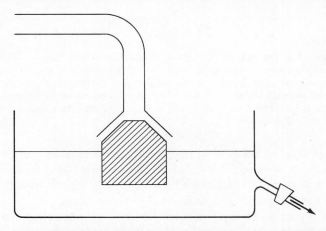

Figure 5.1 The float (shaded) can rise to close off flow from the pipe when the water level is high enough.

pipe feeding water to replace that which may be drawn off through the spigot or otherwise removed. As the water flows in from the pipe and the level rises, the float moves into the opening of the pipe and cuts off the water flow. If the water level drops, the float drops out and more water flows in. In this case, the float serves to sense the condition of too low a water level and also to effect the change in the water level by allowing water to flow. It serves as both the sensor and the effector of the system.

It very often happens that the sensor does not have the power or strength to also serve as the effector. In the case illustrated above, the buoyant force on the float may not be enough to shut off the water flow if the pipe pressure is too high, and water may overflow the vessel. It is then necessary to provide some sort of amplifier. In Figure 5.2(a) we see an example of a float attached to a lever that amplifies the force it exerts on the plug and closes the pipe. This is exactly how the water level is set in a toilet tank.

Figure 5.2(b) shows a similar lever-amplified water level controller that will not work properly. In this case the sign of the response of the system to a

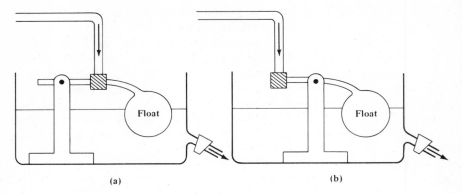

(a)   (b)

Figure 5.2 The float in (a) can rise to cause flow from the pipe to stop when the water level is high enough. The force available to stop the water flow is greater than the buoyant force because of the leverage arrangement. The float in (b) will rise with water level, but because the leverage arrangement is incorrect it will still allow water to flow into the container. The system will not work.

low or high level of water is wrong: it admits water when the level is high and stops the flow when the level is low. In any closed-loop feedback control system it is necessary to make sure that the gain of the amplifier (in this case the lever) has the proper sign to make the system work.

We have seen an example of a simple feedback controlled system in the water tank and now will generalize from it. In order to function properly, a system requries a *sensor*, an *amplifier*, and an *effector*. The sensor must be able to detect the condition of the *controlled variable* and the effector must have the power to maintain it. It would be of little help if the input pipe in the previous example could supply a maximum of 3 liters/minute and water were withdrawn at 4 liters/minute. Also, we have seen that the amplifier must supply a signal of the proper sign to the effector.

A diagrammatic sketch of a simple feedback control system is shown in Figure 5.3. If the controlled variable is perturbed by an external influence, the control system acts to restore it to the desired value. The desired value is often called the *set point* and, in many instances, is adjustable.

We now consider a simple temperature control. The sensor is made by laminating together two pieces of material with different thermal expansion coefficients. When the temperature is increased, the strip will curl toward the side with the smaller thermal expansion coefficient. If one end of this bimetal strip is firmly mounted, the other end can be used to make an electrical contact that operates a heating or cooling device. Such systems are used for regulating the temperature in ovens, toasters, houses, refrigerators, and a variety of other enclosures. In Figure 5.4(a) it is clear that if the strip curls to

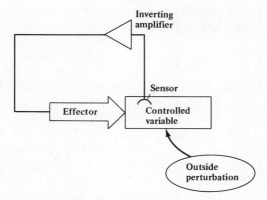

Figure 5.3 Diagrammatic sketch of simple feedback control system.

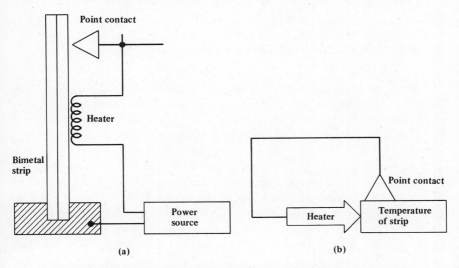

Figure 5.4 Drawing of simple temperature control system of the type found in ovens, toasters, homes, air conditioners, etc. The set point is adjustable by moving the point contact. Part (a) is the actual hardware implementation of the system and (b) is a schematic diagram.

the left when it is heated, it will then operate the heater unit properly and maintain a nearly constant temperature. The set point can be changed by merely moving the contact to the left or right: this is precisely how a home thermostat or toaster control works. In this example, the bimetal strip is the sensor, the heater is the effector, and the contact is the amplifier. If a contact which operated a cooling unit were placed on the left, the device in Figure 5.4 would have greater versatility. A diagrammatic sketch of this system is shown in Figure 5.4(b). Note that movement of the bimetal strip constitutes a signal from the sensor and movement of the contact constitutes a change of the set point.

So far we have restricted the discussion to a particular kind of closed-loop feedback system in which the effector is either on or off. Such an on/off system is to be contrasted with a feedback controlled system that exerts *continuous control*. Various kinds of gas pressure regulators and electronic circuits fall in this latter category. Also, when driving on a curvy road, it is not necessary to have the steering wheel either full left or full right. Instead it is possible to exert continuous control over the direction of the car's travel by turning the wheel to a large number of intermediate positions. It is clear that there are many cases where continuous control systems, in which the input from the effector is proportional to the excursion from the set point, are superior to on/off systems.

Part 2 : Quantitative Description

We shall now begin a quantitative study of feedback controlled systems. One of the most important characteristics of these systems is the inherent *time delay* between the sensor's detection of a deviation from the set point and the effector's response to that condition. In many systems this time delay is very short, but as long as it is not zero (and it can never be zero), it plays a critical role in determining whether or not the system will function properly.

Let us start with several numerical examples beginning with a hypothetical system with gain g and zero time delay. When the excursion of the controlled variable from the set point is A, then the correction applied is $-gA$ and this correction takes place with no time delay. For the first example, we let $g = \frac{3}{4}$, $A = 2$, and we observe the value of A in successive time intervals (see Table 5.1).

TABLE 5.1 *Successive values of the excursion A of the controlled variable from the set point. (Hypothetical case of no time delay)*

A:	2	0.5	0.125	0.031	0.008	0.002
Correction to A:	−1.5	−0.375	−0.094	−0.023	−0.006	−0.002

At the start, $A = 2$ and the correction to be applied is $gA = -1.5$ giving a new value of 0.5 for A. Then a new correction $gA = -0.375$ is to be applied and the resulting value of A is now 0.125. Again there is a new correction of

-0.093 and the resulting value of A is now 0.031. The plot of A versus time in Figure 5.5 shows that it approaches the set point exponentially with time. This very well-behaved system represents the ideal case.

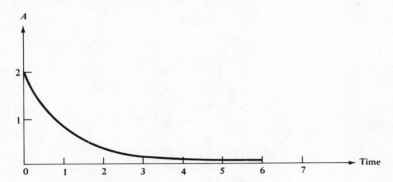

Figure 5.5 Plot of A, the error in the controlled variable, versus time in a simple, proportional control system.

The behavior of a more realistic system with a finite time delay is illustrated in Table 5.2. At the beginning, $A = 2$ but there is no correction to it since it takes a finite time for the system to respond. In the second time interval, the system responds to the value $A = 2$ of the first time interval with a correction of -1.5. The same thing happens during the second time interval and A is driven negative to -1. At this time, however, the system acts as if $A = 0.5$ and delivers a correction of -0.375 driving A to -1.375. At this point the system responds to the value $A = -1$ and delivers a correction 0.75 so that $A = 0.625$. The reader is urged to study the next few steps of the sequence given in Table 5.2. It is clear from Figure 5.6(a), which is a plot of the value of A versus time from Table 5.2, that the system oscillates about the set point and that the oscillation amplitude decreases with time. The qualitative change of behavior from the preceding example, monotonic approach to the set point versus oscillatory approach to it, has been caused by the existence of the finite time delay.

One might guess that the approach to the set point would be hastened if the system had a higher gain. In Table 5.3 we study the same system with the gain increased from $\frac{3}{4}$ to $\frac{5}{4}$. The result, plotted as a solid line in Figure 5.6(b), shows clearly that the oscillating system does not even converge to the set point but instead undergoes oscillations of increasing amplitude until some physical part of the system breaks down. A feedback controlled system WILL NOT WORK IF THE GAIN IS TOO LARGE in the presence of any finite time delay. Furthermore, as the broken curve of Figure 5.6(b) shows, the system undergoes oscillatory divergence if the gain is held constant and the time delay increases.

Table 5.2 *Successive values of the excursion A of the controlled variable from the set point. (Finite time delay).*

A:	2	2	0.5	−1	−1.375	−0.625	0.406	0.875	0.570	−0.086	−0.513	−0.448	−0.065
Correction to A:	0	−1.5	−1.5	−0.375	0.75	1.031	0.469	−0.305	−0.656	−0.427	0.065	0.384	0.336
			0.271	0.319	0.115	−0.124	−0.210	−0.117	0.041	0.129	0.099		
			0.048	−0.203	−0.239	−0.086	0.093	0.158	0.088	−0.030			

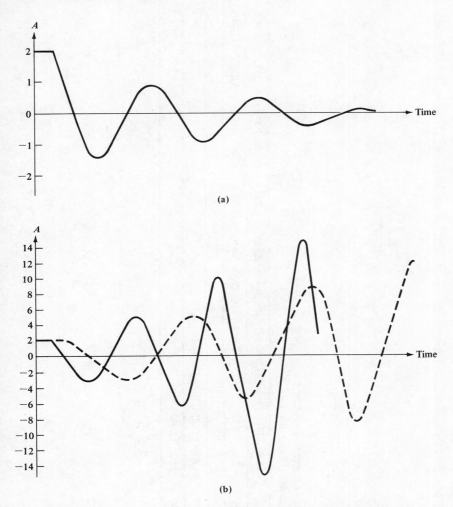

Figure 5.6 (a) The time dependence of the error of the controlled variable A for a proportional control system with finite time delay. The approach to zero error, in the absence of other perturbations, is oscillatory. In (b) the solid curve represents the same system with larger gain but same time delay as in (a); the dashed curve represents the same system with same gain but larger time delay than in (a). Both of these undergo divergent oscillations.

We have seen that the operating characteristics of a feedback controlled system must be carefully chosen in order for it to work properly. We shall now study a more general case. Consider a proportional control system as before with a finite time delay Δt, a gain g (dimensions of 1/time), and a controlled variable p whose set point is p_0. We suppose that during the time

TABLE 5.3 *Successive values of the excursion A of the controlled variable from the set point. (Higher gain.)*

A:	2	2	−0.5	−3	−2.375	1.375	4.344	2.625	−2.805	−6.086	2.580	10.188	6.963
Correction to A:	0	−2.5	−2.5	0.625	3.75	2.969	−1.719	−5.430	−3.281	3.506	7.608	−3.225	−12.74
			−5.77	−14.47	−7.25	10.84	19.90	6.35					
				−8.70	7.21	18.09	9.06	−13.55					

interval from t_2 to $t_2 + dt$ the variable p changes by

$$dp(t_2) = -g\, dt\,[p(t_1) - p_0] \equiv -gA(t_1)\, dt = -gA(t_2 - \Delta t)\, dt \qquad (5.1)$$

where the time delay is manifest in the requirement to evaluate the error signal $A \equiv p - p_0$ at time t_1, which is earlier than t_2 by the amount Δt (see Figure 5.7). Note that $dA = dp$ and find

$$dA(t) = -gA(t - \Delta t)\, dt. \qquad (5.2)$$

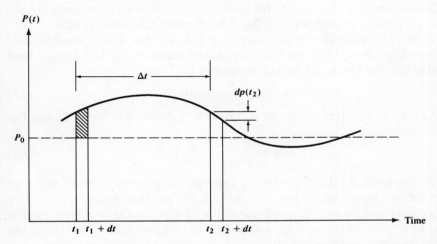

Figure 5.7 A plot of $P(t)$ versus t showing the choices of t_1 and t_2. During the time interval from t_2 to $t_2 + dt$ the system is affected by the total error that existed during the time interval from t_1 to $t_1 + dt$ (shown shaded). The change in P is $dP(t_2)$.

There is no straightforward way to solve this differential equation with a finite time delay, but we might guess that the solution is exponential as it certainly is for the case $\Delta t = 0$. We let

$$A = A_0 e^{zt} \qquad (5.3)$$

where z is some complex number that depends on Δt and g. Substitute Eq. 5.3 into Eq. 5.2 and find

$$z = -g e^{-z\Delta t}. \qquad (5.4)$$

In order to study the solutions of this transcendental equation we shall write $z = x + iy$ and seek the behavior of x and y. Clearly a positive value of x leads to diverging values for A and hence for p so we expect x to be negative for small values of the gain and time delay and positive for larger values. We therefore ask under what conditions $x = 0$. Then z is pure imaginary and Eq. 5.4 gives

$$y = +g \sin y\Delta t \qquad (5.5a)$$

and

$$0 = \cos y\Delta t. \qquad (5.5b)$$

From Eq. 5.5b we find $y = \pi/2\Delta t$ and then Eq. 5.5a becomes

$$g\Delta t = \frac{\pi}{2}. \qquad (5.6)$$

We can see that when $x = 0$, the system will oscillate with amplitude A_0 and at an angular frequency $g = \pi/2\Delta t$, which means the period is $4\Delta t$.

We now consider the case when x is not zero but is very much smaller than $1/\Delta t$. We expand the real part of the exponential in Eq. 5.4 and keep only the first two terms to find, instead of Eq. 5.5a,

$$y = g(1 - x\Delta t) \sin y\Delta t. \qquad (5.7)$$

As long as x is very small, we still have $y \cong \pi/2\Delta t$ as in Eq. 5.5b and Eq. 5.7 becomes

$$x = \frac{1}{\Delta t}\left(1 - \frac{\pi}{2g\Delta t}\right). \qquad (5.8)$$

We can easily see that x is negative and that the system will return to the set point as long as $g\Delta t$ is less than $\pi/2$. If $g\Delta t$ becomes greater than $\pi/2$, x becomes positive and the amplitude of the resulting oscillations diverges. The system loses control of itself and some kind of breakdown will occur unless the stability condition

$$g\Delta t < \frac{\pi}{2} \qquad (5.9)$$

is met. Since the gain and time delay of most systems are not constant but are subject to perturbation by outside influences, it seems desirable to operate well away from the threshold of instability.

The study of the behavior of feedback controlled systems is part of a branch of applied mathematics called *control theory*. We have investigated some of the properties of one solution to one of the simplest problems in this field. It has taught us that controls may oscillate or diverge under a variety of conditions. A more detailed study of these problems is beyond the scope of this book, but suffice it to say that oscillations and divergences are general properties.

Hassenstein has suggested a very interesting demonstration in which it is possible to manipulate the gain of a common control system, the pupillary reflex of the eye. When there is too much light on the retina, the pupillary reflex causes the pupil to close down, and too little light causes it to open. The amount of light on the retina is proportional to the open area of the pupil. Normally, a 10% change in the radius of the pupil results in a 20%

change in its area and in the light level at the retina. If the central part of the pupil is occluded, however, as shown in Figure 5.8, a 10% change in the radius can result in a 100% change in the light level. Thus the differential gain of the system has been altered. It is easy to perform this experiment by cutting a small disc (3-mm diameter) from black tape and sticking it to a small sheet of transparent material (cellophane, plastic sheet, etc.). Hold the sheet very close to one eye with the other closed and try to center the black dot in the pupil while looking at a fluorescent light or other brightly lit large area. Under the right conditions, the pupil will pulsate at about 1 Hz and this is easily observed as a pulsation of the size of the disc. Here is an example of how a normally well-behaved system can be driven into oscillation by an increase in its gain.

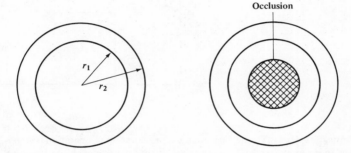

Figure 5.8 Sketch of the pupil of the eye with and without occluding disc.

Part 3: Operational Amplifiers

The widespread use of feedback controlled systems and the detailed study of their behavior has led to feedback stabilized amplifiers and devices. Since the value of the gain g is so important in determining the behavior of a system, it is clearly desirable to stabilize g against the influences of external conditions. We shall study the use of operational amplifiers (op amp) as circuit elements that have to remain very stable.

An op amp is a general term that usually refers to members of a large class of high-impedance, high-gain amplifiers. The open-circuit gain of an op amp is so large that it is generally taken to be infinite, and its input resistance is so high that it is also considered to be infinite. In practice this means that any voltage applied to the input terminals of an op amp will drive the output to the full range of its power supply and that essentially zero current will be drawn from the source producing the voltage. Usually the input stage of an op amp is differential and made with high-gain triodes, electrometer tubes, or field effect transistors (FET). The output stage is a low-impedance follower circuit. Modern op amps are integrated circuits packaged in a single case about the size of an ordinary transistor and may easily contain a dozen transistors and diodes.

The circuit symbol for an op amp is shown in Figure 5.9(a). We shall consider an op amp used in the circuit of Figure 5.9(b) and having a gain α that is very large. R_i and R_f are called the input and feedback resistances, respectively, and V_{in} is the signal from some device that we want to amplify.

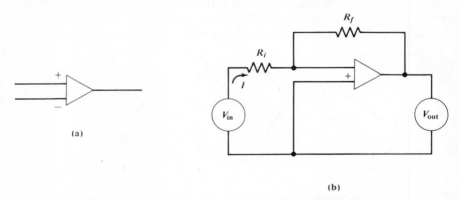

(a)

(b)

Figure 5.9 Part (a) is the diagramatic representation of operational amplifiers in circuits; part (b) shows a typical circuit for a gain-stabilized op amp.

We write Kirchhoff's loop law for the first loop which contains the source, V_{in}, the input stage of the op amp, and R_i. We find

$$V_{in} - IR_i = V \tag{5.10}$$

where V is the voltage across the input terminals of the op amp. Since its gain is α and one of its terminals is grounded, we have $\alpha V = V_{out}$ and Eq. 5.10 becomes

$$I = \frac{V_{in} - V_{out}/\alpha}{R_i}. \tag{5.11}$$

Next we write the loop law for the outside perimeter of the circuit

$$V_{in} - I(R_i + R_f) = V_{out}. \tag{5.12}$$

Here we have required that the current in R_i and the current in R_f be the same because the impedance of the op amp is so high that it draws essentially zero current. We substitute for I using Eq. 5.11, divide both sides of the equation by V_{in}, and define the gain of the circuit g as V_{out}/V_{in}. A little algebra gives

$$g = -\left(\frac{R_f}{R_i}\right)\left[\frac{1}{1 - (1/\alpha)(1 + R_f/R_i)}\right]. \tag{5.13}$$

We notice that, as long as R_f/R_i is chosen small compared with α (and $\alpha \gg 1$), the gain of the circuit is nearly independent of α and is simply

$$g = \frac{-R_f}{R_i}. \qquad (5.14)$$

Since α is usually very large ($\cong 10^5$), it is easy to choose a gain as large as several hundred and have it remain very constant. If the power supplies to the op amp fluctuate, or if α drifts because of temperature changes, or if any other change alters the properties of the op amp, the circuit shown here will maintain its gain at $-R_f/R_i$ with essentially no variation. A change in α by a factor of 2 has almost no effect on g.

The gain of this circuit is said to be feedback stabilized by R_f. Note that in the limit of R_f becoming very large, the gain of the system approaches α, which is the open-circuit gain of the op amp.

The feedback stabilized op amp can be used as an amplifier in a control system where it is required that the gain remain constant. If the output of the thermocouple in some temperature control system is V_{in}, and V_{out} operates a heater, the system can serve as a stable, reliable temperature controller. Furthermore, putting a battery in series with the thermocouple allows one to vary the set point. Feedback stabilization illustrates a principle commonly found in circuit design: one can trade off on one parameter for improvement in another. In this case the gain is reduced but the stability is improved.

Part 4: Body Temperature

The introduction of the previous sections should provide the background for a proper appreciation of some of the control systems of the human body. We shall first examine the temperature control mechanism, which consists of a complicated array of sensors and effectors. It differs from the systems we have discussed so far in that there are several effectors for both raising and lowering the temperatures, and each of them has a different delay time as well as a different capacity for transferring heat. In addition, there are also several sensors of the temperature. Some of these communicate with some of the effectors, and some communicate with all of them. However, all sensors do not communicate with all effectors.

We must first dispense with the idea of a constant body temperature. The *diurnal* (from one part of the day to another) fluctuations of the human temperature are typically about 1°C. Temperature is lowest in the morning after awakening and reaches a peak during the most active part of the day. As activity subsides in the evening, so does the body temperature. It falls further during sleep. These fluctuations are not to be interpreted as a failure of the temperature control system. Rather, they represent variations in the set point of the temperature control mechanism. Even if one measures the temperature at a fixed time every day, there are fluctuations. During the 8 to 10 days between menstruation and ovulation in a normal woman, the body temperature is depressed by about $\frac{1}{4}$°C. Upon ovulation there is a further depression by another $\frac{1}{4}$° for about a day, followed by a rapid rise to normal temperature for the rest of the cycle. The onset of menstruation lowers the temperature again. These changes also derive from variations of the set point. Finally, there are individual differences in any sample of the population. There is a spread about the average 37°C whose standard deviation is about $\frac{1}{2}$°C.

In Figure 5.10 we see some of the many control loops in the temperature regulation system. The primary sensor of the temperature control system is the hypothalamus, a pea-sized organ located in the middle of the brain. There are also sensors in the skin that activate certain regulatory mechanisms.

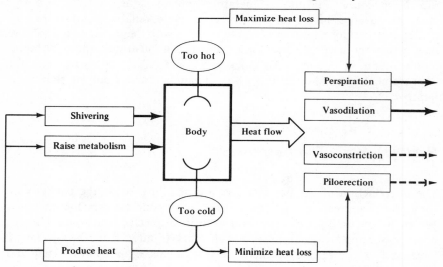

Figure 5.10 Schematic diagram showing the most important control loops for maintaining body temperature in the face of internal and external thermal stress.

Since the core temperature of the human body is almost always kept warmer than its surroundings, the major aspect of the temperature regulating system is simply to vary the rate of flow of heat from the core to the environment.

This is accomplished in a variety of ways. For example, when the skin sensors are activated, they can produce a reflex response we call "goose bumps." These are essentially the result of an erection of the hair follicles (*piloerection*), which, in hairy animals, would result in an effective thickening of the furry layer. The increased insulating depth serves to restrict heat flow as we have seen in Chapter 2.

In order for heat to be lost to the surroundings, it must be brought to the surface. Of course it can be conducted there, but it is far more efficient to convect it using the blood as a carrier. We have seen in Chapter 2 that vasodilation serves to bring more blood closer to the surface and thus frustrate the heat conserving process of countercurrent exchange. The opposite process, vasoconstriction, serves to reduce heat exchange with the environment. These processes are responsible for the "flushed" appearance of someone who is overheated and the pale color of someone who is chilled.

Another way for the body to give up heat to the environment is through perspiration. Perspiration serves to reduce body temperature by carrying away heat when the water evaporates. Unlike the mechanisms discussed above, this one works when the surrounding temperature is greater than body temperature. In fact, at constant relative humidity, the cooling effectiveness of perspiration increases as the temperature does because the water evaporates faster. Of course it is necessary to replace the lost water or the body's fluid balance will be destroyed.

In addition to conserving heat by piloerection and vasoconstriction, the body can warm itself by increasing its rate of heat production. This can be done by an increase in metabolic rate by the organs in the core or by an increase in energy consumption and heat production by the skeletal muscles. This latter process is accomplished by rapid muscular contraction and relaxation, which we call "shivering." Since 70 to 80% of the energy consumed by muscle ends up as heat and since shivering produces little external work so that most of the remaining energy also goes into heat, it is an effective mechanism for warming the peripheral areas. The sensor and effector of the shivering mechanism are unusually well matched because they are triggered by a rapid drop of skin temperature and the mechanism heats the peripheral areas.

When an individual contracts certain diseases, the set point of his body temperature is rapidly elevated. He feels cold and may even shiver in order to raise the temperature to the feverish set point. Even as the temperature rises considerably above normal, the need for further warmth will be manifest in shivering and pale appearance. These symptoms will disappear only after the fever has been reached and the temperature is high. After some time the body's defense will defeat the disease and the set point will return to normal. The body will then seek ways to eliminate the surplus heat energy and the patient will be flushed and sweat profusely until the temperature comes down.

The elevated temperature associated with certain diseases is one of the body's mechanisms for fighting the disease and in general we should not interfere. Taking large doses of aspirin to control a mild fever may very well prolong the illness. The detrimental effects of controlling the body temperature during an infection have been well documented in laboratory animals. Of course, the cells of the brain, liver, and other organs can be irreparably damaged by a sustained high fever ($41°C \cong 105°F$), especially in small children. Such a high fever is an overreaction of the feedback control system to the sudden change in the set point, and it should obviously be opposed with cold baths and aspirin. There are many old wives' tales about proper treatment of patients during these periods of rising and falling temperatures: it's obvious that most of them should be ignored.

Part 5 :   Other Feedback Controlled Systems

Perhaps the most highly developed feedback control system in the human body is responsible for control of motion. The simple act of reaching for a nearby object and the highly complex motions of the aerial acrobat are governed by a complex control system involving several sensors and effectors. Muscular motion is not characterized by a single set point that is to be maintained in the face of a changing environment. Instead, there is a series of changing set points (positions) that must be achieved in sequence.

Consider the act of writing. The fingers move the pen in a complicated path to produce the individual letters while the arm and shoulder move the hand across the page. At the same time, the pressure on the pen is varied to allow for spaces between words, and, in the days of feather quills, width of line. The sensor in the system is the eye. We look at the position of the pen and consciously correct for its deviations from the desired positions as it moves across the page. The amplifier(s) are the muscles of the fingers, hands, and arms that receive millivolt impulses from the nerves and produce motions of centimeters and forces of thousands of dynes. The effectors are the bones and tendons that move the pen from place to place according to command. The result of all this complicated interaction is a line or page of text that we normally take for granted.

It should be noted that it is quite possible to write with closed eyes. If the visual system is an essential part of the feedback loop, how is it possible to do this? It turns out that most of the muscles in our bodies are equipped with strain gauges that telemeter information from the muscles to the central nervous system; that is, the brain has information about the position of the muscles and, hence, the limbs. When certain sequences of muscular motion have been performed many times under the guidance of the feedback control system described above, it is no longer necessary to use the visual feedback—the muscular feedback provides adequate information. For example, most people can tie their shoes without looking. Obviously a different task (tying a new knot) or a precise one (printing block letters) requires the surveillance of the visual system for adequate feedback.

There is a large class of well-developed systems for regulation of the levels of certain chemicals in the body. These include hormones, blood sugar, blood saline, certain metallic ions (e.g., calcium), proteins, as well as certain dissolved gases and pH. There may be several dozen separate or partially interacting feedback control loops for chemical stabilization, but we shall discuss only the one for regulating calcium.

Calcium concentration plays a role in many different body functions. It plays an important role in the behavior of excitable membrances, it is particularly important to muscle contraction (especially heart muscle), and it is an important structural constituent in bone. In fact, there is so much calcium in the bones, about 2% of body weight, that the skeleton serves as a reservoir for the minute amounts of calcium needed for proper nerve and muscle function. Plasma calcium levels are normally maintained between 85 and 105 micrograms/milliliter. The total amount of calcium in the blood is therefore only a gram or two in normal circumstances.

Calcium may enter the blood plasma directly from the food processed by the digestive system or it may be drawn in from the large skeletal reservoir. Calcium may leave the blood plasma through urinary excretion or by deposition into the bone. It may also be rendered temporarily inactive by being bound to proteins. The protein-bound calcium thus forms another reservoir from which it may be drawn when needed.

The control system for calcium concentration operates by regulating the flow in and out of the body as well as in and out of the reservoirs. In that sense, its schematic diagram looks very much like the temperature control mechanism. There are several ways to add it to the bloodstream as well as remove it from the bloodstream just as there are several ways to add or remove heat.

The sensor for this system is located in the parathyroid glands that secrete a parathyroid hormone. This hormone acts to draw calcium from the intestinal walls if it is present, to draw it from the bones, and to inhibit its excretion by the kidneys by causing more of it to be reabsorbed in the renal tubules. When the blood calcium exceeds 105 micrograms/milliliter, the thyroid secretes a hormone called *thyrocalcitonin* that increases the net entry of calcium into the bones and thereby holds down the concentration.

The effectors for calcium maintenance are therefore many and varied, and they must somehow act in concert to prevent large fluctuations of concentration. On the other hand, it is one of the simpler systems because the concentration does not vary much (fixed set point) and the feedback loop does not involve brain function.

There are many other feedback controlled systems in the human body that cannot be discussed here. Some of the mechanical ones are regulation of blood pressure, muscle tension, pupil diameter, balance, and lens focus. In addition there are several that are physicochemical in nature such as blood osmolarity, respiration rate, and heart rate. Also, there is the largest group of purely chemical (biochemical) systems mentioned above. Finally, there are some special systems that do not fall easily into any of these classes. The complicated genetic regulation process for replication of DNA is the prime example of such a controlled process.

We have neglected any discussion of noise in these systems because it is a bit complicated to deal with. Noise is the injection of unwarranted (and unwanted) information or signals in the sensor, effector, amplifier, or other parts of the control loop. Most systems are adequately protected against the influence of noise by a variety of mechanisms. The easiest protection scheme is simply temporal filtering: most noise is in the form of short disturbances and most control systems have time constants that are longer than the typical times of the noise disturbance. Another protection mechanism depends on extreme specificity to a particular signal, e.g., a chemoreceptor on a dendrite. There are many other schemes as well.

Questions and Problems for Chapter 5

1. A feedback controlled system for maintaining the level of water in a tank consists of a 5-cm cubic wooden float (density $= 0.7$ g/cm^3) that floats up against a 1-cm diameter pipe to shut off the flow when the water reaches the proper level. What is the maximum allowable pressure in the supply pipe?

2. Be sure you can draw a schematic diagram of a feedback controlled system and label and explain the function of all the major parts.

3. It is said that a human cannot survive on seawater because the kidneys cannot extract and excrete enough salt from it. On the other hand, the kidneys are the effector in a feedback controlled system to maintain the salinity of the blood. Explain the apparent contradiction and give an example of another feedback controlled system that can encounter a similar circumstance.

4. Explain how it is that a system under negative feedback control can oscillate wildly. Give an example.

5. Why is it that, at the onset of a fever, a patient may feel chilled even though his body temperature may be several degrees above normal?

Bibliography

1. LOREN CARLSON and ARNOLD HSIEH, *Control of Energy Exchange* (Macmillan Publishing Co., Inc., New York, 1970). (A detailed, quantitative treatment of energy balance in the body.)

2. B. HASSENSTEIN, *Information and Control in the Living Organism* (Chapman and Hall, London, and Barnes and Noble Books, New York, 1971). (An interesting discussion of a variety of topics.)

3. R. HOBBIE, *Intermediate Physics for Medicine and Biology* (John Wiley & Sons, New York, 1978.) (Discussion of systems with multiple time delays.)

4. OTTO MAYR, "The Origins of Feedback Control," *Scientific American Offprint No. 338* (October 1970).

5. DOUGLAS RIGGS, *Control Theory and Physiological Feedback Mechanisms* (The Williams and Wilkins Company, Baltimore, Md., 1970). (A rigorous, mathematical discussion.)

6. E. SELKURT, ed., *Physiology* (Little, Brown and Company, Boston, Mass., 1971). (A standard physiology text.)

7. C. L. STRONG, "The Amateur Scientist," *Scientific American* (January 1971). (A description of op amps and their uses in many circuits.)

8. JOHN TRUXAL and LUDWIG BRAUN, "Bioengineering: The Body as a Machine," *Impact of Science on Society*, Vol. XX, No. 4 (1970).

9. ARNOLD TUSTIN, "Feedback," *Scientific American Offprint No. 327* (September 1952).

10. RICHARD WURTMAN, "On Teaching How the Body Works," *Educational Research Center of Massachusetts Institute of Technology* (1969).

6

NERVE CELLS

Part 1 : Observations on the Resting Axon

Our study of the nervous system will be restricted to certain simple electrical properties of nerve cells. We shall not study many of the problems of chemical transmitters or membrane properties that are two of the most active areas of biological research today. This does not mean that these subjects are of little importance; however, their proper study would be the subject of another book.

Nerve cells are different from any others in the body. All cells have a main cell body with a nucleus, but nerve cells have a long tube called an *axon* extending from the cell body (see Figure 6.1). In the human body there are nerve cells that are so long that their cell body may be in the spine and the axon may terminate in the foot. At the end of the long axon are a variety of hairlike structures for communicating information to muscles or other

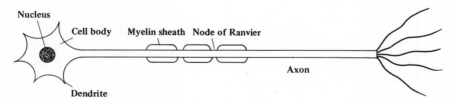

Figure 6.1 Drawing of a nerve cell showing major parts.

nerve cells. In some cases the axon may be wrapped in a series of small *sheaths of myelin* separated every few millimeters by nodes (*node of Ranvier*). These myelin sheaths result from the spiral growth of Schwann cells around the axon. We shall devote most of our attention to the axon, and particularly its electrical behavior. We shall see that axons act as electrical communication lines carrying information throughout the body.

The membrane of the axon separates the inner fluid (axoplasm) from the outer fluid. It is made of a double layer of strongly polar phospholipid molecules about 100 Å thick and has a dielectric constant of about 5. Figure 6.2 shows that these molecules are arranged with their hydrocarbon tails on the inside and polar lipid ends forming the boundary of the membrane. The outer boundary often has many protein molecules located on or penetrating its surface.

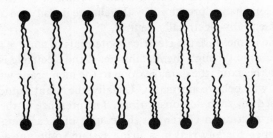

Figure 6.2 Alignment of phospholipid molecules to form membrane.

In the normal resting state, the inside of a nerve axon is about 70 mV negative with respect to the outside (see Figure 6.3). Since the membrane thickness is very small compared with the axon radius, we can approximate the axon membrane as a plane parallel plate capacitor that has been rolled into a tube. The charge Q is given by $Q = CV$ and the capacitance $C = A\epsilon/t$ where A is the area and t is the thickness. Here $\epsilon = 5\epsilon_0 = 5 \times 10^{-10}/4\pi \cong 5 \times 10^{-11}$ farad/m. For $t \cong 100\,\text{Å} = 10^{-8}$ m, we find a capacitance of 5×10^{-3} farad for each square meter of membrane or $\frac{1}{2}$ microfarad/cm^2.

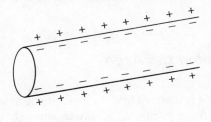

Axon

Figure 6.3 Sketch of axon showing charge distribution at rest.

Since we know the voltage on this capacitor is typically 70 mV, the charge per square centimeter is 3×10^{-8} coulomb. This charge is produced by separating the ions in the axoplasm.

The concentration of monovalent ions present in a typical body fluid is about 150 millimoles/liter, which means that a cubic centimeter carries 1.5 $\times 10^{-4}$ mole or about $1.5 \times 10^{-4} \times (6 \times 10^{23}$ charges/mole$) \times (1.6 \times 10^{-19}$ coulomb/charge$) = 15$ coulombs of both positive and negative charge. Since the amount of charge needed for a square centimeter of the membrane is only 3×10^{-8} coulomb, it is clear that the membrane needs the ions from a layer only $\frac{1}{5}$ Å thick to produce the 70 mV resting potential. A typical ion is about 1 Å in diameter, which means that only one-fifth of those ions in the single monolayer adjacent to the membrane need to be transported across it in order to produce the resting potential. Since the diameter of an axon is 10,000 Å or more, this layer is only a small fraction of the total volume and the concentration change produced by the charge density is so small that the axoplasm is essentially electrically neutral.

It is possible to measure the electrical potentials in a nerve axon by inserting a microelectrode through the membrane. If the electrode is small enough compared with the axon, the damage done to the membrane will be miniscule and the axon will behave normally. Usually the giant squid axon (1-mm diameter) is used for such demonstrations. The presence of the probe creates problems when it draws a current in the course of measuring the potential, which will happen for any real measuring system (only an ideal system can have infinite resistance and zero stray capacitance). If the current is negative, K^+ and Na^+ ions will be neutralized and will then react with the water (electrolysis) upsetting the chemical balance and pH of the axoplasm. For this reason, the electrodes are usually made of silver and coated with AgCl. When the current is negative, Cl^- ions go into solution; when it is positive, they plate out in the form of AgCl. Since a very common negative ion in the solution is Cl^-, there is little change in the axoplasm chemistry. Such electrodes are called *reversible*. They provide an interface between the electron-carried current in the metal probe and the ion-carried current in the axoplasm. The electrical resistance of these electrodes is high, but at least it's the same in both directions.

Part 2 : Perturbing the Axon

When a microelectrode is inserted into an axon, its observed potential drops suddenly to the resting -70 mV as it passes through the membrane. In myelinated axons the drop is not as sudden as it is in unmyelinated axons because of the insulating property of the myelin. We ask what happens to an axon if a voltage is applied to an electrode that has been inserted into the cell; that is, we no longer consider a passive experiment but now try to change the potentials in the axon.

We choose a model for the axon with a resistivity ρ_a for the axoplasm and a much larger ρ_m for the membrane (the capacitance is about $0.5\ \mu\text{F}/\text{cm}^2$ as before). We assume the axon is bathed in a fluid whose resistivity is comparable to that of the axoplasm, but we note that the outside fluid has considerable spatial extent. Therefore the resistance to current flow in the bath is much less than in the axon. We therefore assume the bath to be an equipotential volume and choose it as electrical ground.

When an electrode has been inserted through the membrane, it is at a potential of -70 mV relative to ground. If we impress upon it a different voltage, say -60 mV, a current will flow. In order to find the distribution of voltage resulting from the -60-mV electrode we shall calculate the current distribution. Consider a disc-shaped slice of the axon located at coordinate x and of thickness dx as in Figure 6.4. We choose the origin of coordinates at the electrode: $V(0) = -60$ mV. Then

$$V(x) - V(x + dx) = I_a R_a = \frac{I_a \rho_a\, dx}{\pi r^2} \tag{6.1}$$

where I_a is the axon current and depends on x, and r is the radius of the axon. The left-hand side of Eq. 6.1 is simply $-dV$ whence

$$\frac{dV}{dx} = \frac{-I_a \rho_a}{\pi r^2}. \tag{6.2}$$

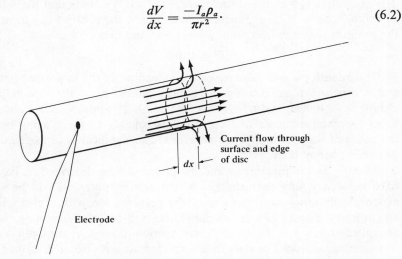

Current flow through surface and edge of disc

dx

Electrode

Figure 6.4 Arrows show current flow at a particular point. All the charge that flows into the disc must flow out either through the edge (axon membrane) or through the plane boundary, which means it flows further along the axon.

In order to conserve electric charge, any current flowing out of this disc through the membrane dI_m must be equal to the difference between the current entering and leaving the disc through its faces. We write

$$dI_m = I_a(x) - I_a(x + dx) = -dI_a \qquad (6.3)$$

which tells us that

$$\frac{dI_m}{dx} = -\frac{dI_a}{dx}. \qquad (6.4)$$

We use the resistivity ρ_m of the membrane to find the resistance to current flow R_m across a small, plane area and then find the voltage across the membrane from the current flowing across the boundary of the edge of the disc. It is

$$V = dI_m R_m = (dI_m)\left(\frac{\rho_m t}{2\pi r \, dx}\right) \qquad (6.5)$$

where t is the membrane thickness and $(\rho_m t / 2\pi r \, dx)$ is the resistance. From this we obtain

$$\frac{dI_m}{dx} = \frac{2\pi r V}{\rho_m t} = -\frac{dI_a}{dx}. \qquad (6.6)$$

We now differentiate Eq. 6.2 with respect to x and substitute Eq. 6.6 to find

$$\frac{d^2 V}{dx^2} = \frac{2\rho_a}{\rho_m r t} V \equiv \frac{V}{\lambda^2}. \qquad (6.7)$$

The quantity λ is a constant (independent of x). We know that the solutions to Eq. 6.7 can only be exponential. We discard the positive exponential on the grounds that it makes no physical sense and find

$$V = V(0)e^{-x/\lambda}. \qquad (6.8)$$

The quantity λ is a measure of the spreading of the impressed potential and it can be evaluated using the typical values: $\rho_m = 10^8$ ohm-m, $\rho_a = 2$ ohm-m, $r = 10^{-6}$m, and $t = 10^{-8}$m. We find $\lambda \cong \frac{1}{2}$ mm, which is about 500 times the axon diameter and corresponds roughly to the spacing between the nodes of Ranvier in myelinated axons. The significance of this result will become apparent later.

As long as the membrane and axoplasm can be described as having a fixed resistivity, any perturbation of the resting potential will be spread exponentially away from the site of the perturbation with a characteristic length λ in the order of a millimeter. These well-behaved spreading voltages are called *graded potentials* and the conduction of them by the axon is called *saltatory conduction*. The data from experiments with isolated axons fit Eq. 6.8 very well and yield appropriate values for λ over a rather large range of applied voltages.

There is a threshold voltage, however, that completely changes the resistive character of the membrane. In Figure 6.5 we see a plot of the graded

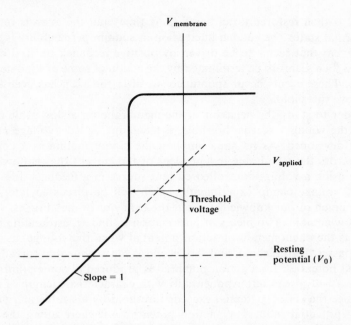

Figure 6.5 Plot of membrane potential versus applied voltage. The membrane behaves like a normal insulator in the region where the slope = 1 but initiates an action potential above threshold.

potential at the site of the electrode versus the applied voltage. For all reasonable voltages below resting potential, and for 20 or 30 mV above it, the graded potential at the origin equals the applied voltage. As the voltage is brought above a certain value, however, the membrane suddenly conducts positive ions into the cell, and the voltage shoots sharply upward. This sudden rise is called an *action potential* and is characteristic of excitable cell membranes.

The behavior of the membrane potential after it has been pushed beyond threshold is a very complicated matter. The action potential, which can go as high as 50 mV positive, derives from a sudden influx of sodium ions that are normally at very low concentration in the axon. Even though the total current that flows is large enough to reverse the charge on the membrane, it is still a very small amount of charge compared with the amount available in the solution. This means that the sodium concentration changes by a miniscule amount and that the axoplasm still remains essentially electrically neutral.

The sodium influx occurs because of a sudden change in the permeability of the membrane to sodium and is followed by a comparably sudden restoration of near impermeability. For reasons we shall discuss later, the resting potential is then restored (after some short time) and the axon is returned to its normal state. The sudden fluctuation in sodium permeability is a self-feeding event that seems to be driven by positive feedback up to a certain point and then seems to be terminated by the action of some of kind negative feedback. These events occur spontaneously after the axon has been stimulated above threshold.

In order to study the behavior of the membrane on a slow scale during the time the voltage exceeds threshold, a technique called "voltage clamp" has been developed. As its name implies, the circuit maintains a constant voltage across the membrane independent of the current that is flowing. It does this using a voltage-controlled current source in a feedback loop. The results of voltage clamp experiments, which will be discussed later, have provided much of our knowledge about the behavior of membranes.

We now combine two pieces of information to find an astounding result. The first is the phenomenon of action potential when the voltage inside the axon surpasses the threshold, and the second is the exponential spreading of the graded potential. If an action potential is produced at some point along the axon, it will spread out exponentially with characteristic length of about 1 mm. Since the action potential exceeds threshold by a sizable margin, the spreading potential induces an action potential elsewhere along the axon (see Figure 6.6). This induces a further action potential and the result is that

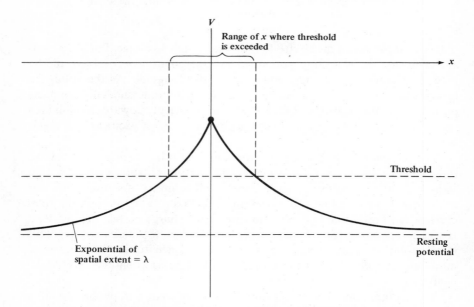

Figure 6.6 Plot of the spatial variation of membrane voltage for a voltage perturbation. Action potential is omitted.

action potentials can propagate along an axon! This astounding behavior occurs as a result of the membrane suddenly conducting and allowing itself to become depolarized from its resting state.

The calculations of the electrical properties of axons in these two sections are called *cable theory*. Cable theory has been used to describe a much larger variety of phenomena than those above. Matsumoto and Tasaki have calculated the velocity of propagation of action potentials in unmyelinated axons and found $v = \sqrt{d/8R^*\rho_a C^2}$ where d is the axon diameter, C is the capacitance per/unit area, and R^* is the resistance of an excited area of membrane.

Part 3: Temporal Characteristic of Axons

So far we have considered the electrical properties of axons under dc excitation. We now consider the time-dependent character of an axon's response to a stimulus. We examine the behavior of the axon when an electrical pulse instead of a dc level change is used to stimulate the axon. If the pulse is very long, it is clear that the axon will perceive it as a level change and the behavior will be unchanged from that of our earlier description. We might guess that if the pulse is extremely fast, the axon will not have time to respond and will therefore ignore it. We shall study the intermediate time region where the pulse is on for a long enough time so that the current from it can appreciably affect the charge on the membrane and therefore change the voltage.

We model the membrane as a parallel *RC* circuit as shown in Figure 6.7. The capacitance and resistance are those of the membrane, and the battery is the source of the resting potential. The capacitance would lose all its charge through the resistance if the battery voltage were zero. When the voltage on an electrode inserted into the axon is suddenly pulsed, it's as if the battery voltage in our model circuit were suddenly pulsed from its resting

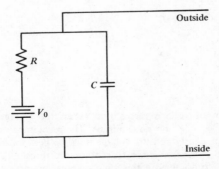

Figure 6.7 Circuit model of a membrane.

value V_0 to a new value V_0' and back to V_0. We apply Kirchhoff's law to the loop in Figure 6.7 during the pulse and find

$$V_0' - IR = V \tag{6.9}$$

where I is the time-dependent current after the pulse is applied and V is the time dependent capacitor (membrane) voltage. Since the charge on the membrane $Q = CV$ and $I = dQ/dt$, we find Eq. 6.9 becomes

$$\frac{dQ}{dt} + \frac{Q}{RC} - \frac{V_0'}{R} = 0. \tag{6.10}$$

The boundary conditions are $Q = CV_0$ at $t = 0$ (before the pulse begins) and $Q = CV_0'$ after very long times (level change). Then the solution to this differential equation is

$$Q(t) = CV_0' + C(V_0 - V_0')e^{-t/RC} \tag{6.11}$$

and $V(t) = Q/C$ is plotted in Figure 6.8.

We see that the membrane potential changes exponentially from V_0 to V_0'. If V_0' is greater than threshold, an action potential will always be generated after a long enough time. If V_0' is considerably greater than threshold,

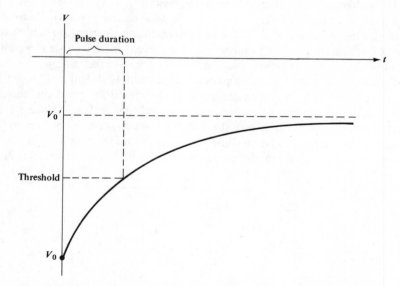

Figure 6.8 Plot of voltage versus time when the voltage is suddenly changed from V_0 to V_0' on a membrane. The membrane doesn't change voltage instantaneously and the voltage must be held for at least a time interval labeled *pulse duration* in order to exceed threshold. If it changes from V_0' back to V_0 before this time elapses, no action potential is generated. Action potential is not shown for $V >$ threshold.

the action potential will be generated in a short time. In fact, for times small when compared to RC we can expand the exponential in Eq. 6.11 and find

$$V \cong V_0 - (V_0 - V_0')\frac{t}{RC}. \tag{6.12}$$

From Eq. 6.12 we see that V can reach threshold V_t when the product of the pulse height $V_0' - V_0$ and the pulse duration t equals or exceeds the constant $(V_t - V_0)RC$. This defines a reciprocity relationship between the pulse length and pulse height required to initiate an action potential. We use the expression $R = \rho_m t/2\pi r dx$ (see Eq. 6.5) for the resistance and $C = A\epsilon/t = 2\pi r\, dx\epsilon/t$ for the capacitance of a unit length of membrane to find

$$RC = \rho_m\epsilon \cong 5 \text{ milliseconds} \tag{6.13}$$

independent of length. Then the reciprocity condition becomes

$$\text{pulse length (msec)} \times \text{pulse height (mV)} > RC(V_t - V_0) \cong 100 \tag{6.14}$$

since V_t is typically -50 mV and $V_0 \cong -70$ mV. The plot of this relation in Figure 6.9 shows that the reciprocity fails for long times because the action potential obviously cannot be started by a dc voltage of any duration if it doesn't exceed threshold.

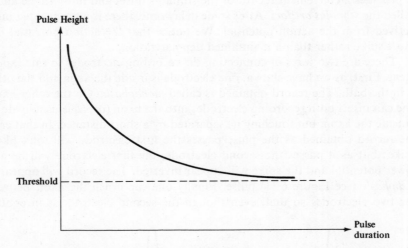

Figure 6.9 Plot of amplitude of voltage pulse versus its duration for pulses that just exceed membrane threshold.

Experiments on isolated, excised single nerve axons are usually performed in a proper bath. A microelectrode is inserted in one end to provide a stimulus pulse and in the other end to record the transmitted nerve impulse. The electrical ground for both stimulating and recording circuits is the

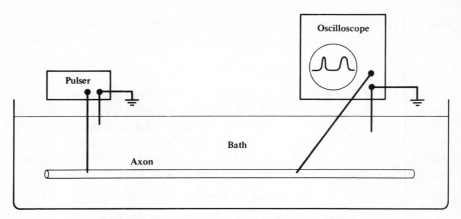

Figure 6.10 Sketch of apparatus used for studying axons.

extracellular fluid. The temporal course of the nerve impulse is displayed on an oscilloscope as shown in Figure 6.10.

If the stimulus is a short pulse ($\cong$ 1 msec), the oscilloscope shows a trace like the one in Figure 6.10. The first pulse is a result of direct-current leakage through the axoplasm and is essentially the same as the stimulus. It provides a convenient record of the stimulus onset and pulse shape and is called the *stimulus artifact*. After some delay time, there is a propagated pulse derived from the action potential. We notice that the action potential is a short spike rather than a maintained depolarization.

There are two ways of connecting the recording electrodes in this experiment. First, as we have shown, one electrode is inside the axon and the other is in the bath. The record obtained is called *monophasic*. On the other hand, one can insert both recording electrodes into the axon (or, as is usually done, outside the axon, but touching it) separated by a short distance. In that case, the record obtained as the pulse passes the first electrode will look like a spike; but as it passes the second electrode, the first electrode will be at a lower potential and the spike will appear inverted. The record will appear as a *diphasic* (see Figure 6.11) pulse. Finally one can crush the axon between the two electrodes so that, even though the second electrode is in contact

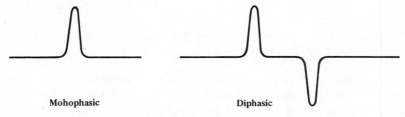

Figure 6.11 Monophasic and diphasic pulses conducted by axons.

with the axoplasm, it will not see the action potential. This provides a mono-phasic record. We now list some results from this experiment.

1. There is a certain elapsed time between the stimulus artifact and the arrival of the spike at the recording electrode. When this is divided into the distance between electrodes, the propagation speed results. Typical values are several meters per/second. Speed is larger for larger axons and for myelinated axons.

2. We can determine what happens if one stimulus is followed by another after a very short time. We find that action potential spikes will be propagated only if a fixed minimum time delay is allowed between stimuli. During this delay time, the axon is essentially passive and will not propagate an action potential. If we decrease the voltage of the second pulse, we find that the time delay, or *refractory period*, is lengthened. As long as the weakened second pulse remains above threshold, however, there will be a propagated action potential after a sufficiently long delay between pulses. Similarly, an increased voltage results in a shorter refractory time. There is a minimum refractory period that cannot be further shortened by increasing the second pulse height further. This is an *absolute refractory time* (typically 1 msec) as compared with the *relative refractory time* that varies with pulse height.

   The existence of this refractory interval has an interesting and important consequence. As soon as an action potential is initiated at one point on an axon, say $x = 0$, the graded potential spreads along the axon initiating action potentials as far away as $\pm \lambda$. The newly initiated action potential, of course, spreads its graded potential and can propagate in both directions, but a propagating action potential at any point on the axon CANNOT initiate a backward propagating action potential. This is because the section of the axon that has just been traversed by the action potential is now refractory and cannot be excited again for some time interval. By the time this interval has elapsed, the action potential has propagated so far away that its graded potential is below threshold. If the refractory behavior of the axon did not exist, a depolarization anywhere along the axon would result in a chaotic and random distribution of action potentials.

3. There exists a maximum rate of repetition of action potentials. This is determined by the refractory period: clearly the action potentials must be separated by an interval at least as great as the absolute refractory time.

4. The height of the action potential is determined by the electrochemical properties of the solutions inside and outside of the axon. Therefore, the height of a pulse is independent of the size of the stimulus that

initiated it and all of them are therefore the same height! This is sometimes referred to as *all or none principle.*

5. In order to propagate the spike, the graded potential must spread along
the axon to exceed threshold somewhere else. The further it spreads,
the faster the action potential will propagate because graded potentials
spread very fast, and the rise time of the voltage change associated
with Na influx is independent of position on the membrane. Since the
myelin sheath on some axons has a very high resistance, the distance
λ that characterizes the exponential spread is much larger in myelinated
axons. Therefore, they conduct at a much higher speed than unmyelinated axons. The myelin also serves to insulate adjacent axons from
one another so that cross talk between axons in a nerve bundle can
be minimized.

The problem with insulated axons is that the spatial exponential decay
of the voltage would limit the distance over which a spike can be transmitted
to a few millimeters or 1 cm because the ions in the extracellular fluid can't
penetrate the axon to conduct the action potential. The myelin sheath is not
continuous, however, and Na ion current flows into the axon at each break
in the sheath (nodes of Ranvier). At each node, the full action potential is
realized. This is because the graded potential spreads with a characteristic
length large enough to allow threshold to be exceeded at that distance. The
nodes serve as little amplifying stations or repeating stations and keep the
pulse propagating along the entire length of the axon.

The entire preceding discussion has been centered around the properties
of a single axon. Most nerves consist of a large bundle of axons, and the
properties of the bundle are different. Although each axon in a bundle may
operate according to an all or none principle, the entire bundle may exhibit
a whole range of responses depending on the stimulus strength. This range
of pulse heights derives simply from the number of axons in the bundle that
are excited. Also, since axons of different sizes propagate the spikes at different speeds, a sharp stimulus may result in a transmitted pulse that is
smeared in time or otherwise has its shape changed. There are also other
differences.

Part 4: The Origin of Membrane Potentials

If we wish to understand the electrical behavior of excitable cells, it is necessary to study the chemistry that governs the production of the potentials. We
start with certain assumptions or postulates that must be supported by the
final description in some self-consistent way.

1. All voltages and changes therein result from the properties of the membranes that separate salt solutions of different concentrations.
2. Active transport is therefore not directly the major source of the potentials.
3. Voltages are all passive diffusion potentials that result from concentration differences. Under normal conditions these can only be changed by changes in membrane permeability.

We consider a container divided by a membrane (see Figure 6.12) whose permeability is not the same for positive and negative ions. In the simplest

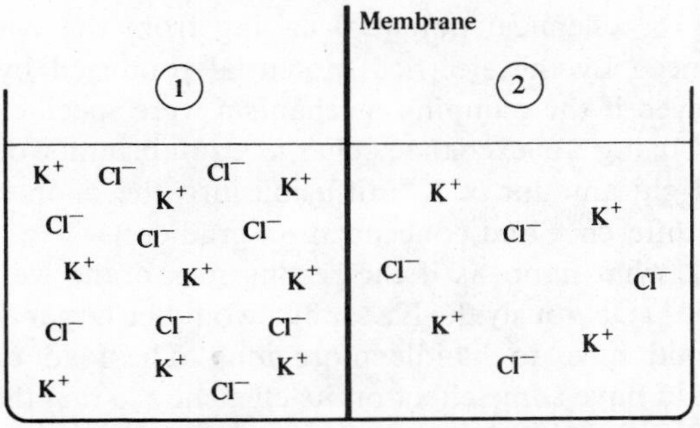

Figure 6.12 Vessel divided by semipermeable membrane for observing electrochemical potentials.

case, suppose it were only permeable to positive ions, K^+. If we prepare the container with two solutions of differing concentrations of KCl on either side, then K^+ ions begin to diffuse immediately from the higher concentration to the lower. Since the Cl^- ions are held back, the two solutions immediately acquire opposite charges that accumulate at the boundary and the resulting voltage difference quickly stops the flow of K^+ ions. Of course, exchange will continue, but the net migration of K^+ through the membrane is stopped.

The voltage V that has built up across the membrane is determined by the chemical potential associated with the concentration difference of the ions. It is given by

$$V = \left(\frac{RT}{zF}\right) \ln \frac{[K_1^+]}{[K_2^+]}, \tag{6.15}$$

where the square brackets indicate the concentration on sides 1 and 2 of the membrane, R is the gas constant, T is the Kelvin temperature, z is the valence (usually $z = 1$), and $F = 9.65 \times 10^4$ coul/mole is the Faraday constant. We note that, at room temperature, RT/zF is about 26 mV. Since resting potentials are typically 70 mV, the log term is about 2 or 3, which means that concentrations do not differ by more than about a factor of 10. Equation 6.15

is called the Nernst equation and appears fairly frequently in physical chemistry.

With this ideal model, we now try to examine a real membrane. We find that no permeability is exactly zero. In fact, the resting permeability to Na^+ is typically 75 times less than it is for K^+. Experiments have established that there is a metabolically supported sodium-potassium pump that slowly removes Na^+ ions from inside the axon and replaces them with potassium. The K^+ ions, which are in greater concentration inside than they are outside, begin to diffuse out, but the net charge transfer cannot be balanced by the return of Na^+ ions because the membrane is much less permeable to them. The electrical potential appears, just as in the example above, and stops the K^+ diffusion when the chemical potential arising from the concentration difference is balanced by the electrical potential produced by the charge. It turns out that even if the pumping mechanism were specific to sodium ejection instead of being an exchange, charge equilibrium would provide for essentially the right amount of K^+ diffusion into the axon to maintain the same potential difference and concentration gradient.

We now ask what happens if the resting membrane were slightly more permeable to Na^+. Obviously the Na^+ influx would be larger and the pumping mechanism would have to handle more ions. The load on the pumping mechanism would have some effect on its efficiency so that the K^+ concentration inside the cell would be lower and the polarization across the membrane would be slightly reduced. The system would reach a new equilibrium and would maintain the new polarization as long as it was not further perturbed.

If the permeability to Na^+ were suddenly increased substantially so that the Na^+ influx became greater than the potassium efflux, the polarization of the cell would decrease to zero and even reverse itself. It is important to realize that this process is not simply self-sustaining but that the process accelerates itself because as the membrane becomes depolarized, its permeability to Na^+ increases encouraging further depolarization via Na^+ influx. The permeability of the membrane to Na^+ quickly becomes 20 to 40 times greater than the permeability to K^+, so K^+ diffusion may be neglected. It is not known why, after about 1 msec, the runaway positive feedback behavior is suddenly stopped and the affected region of the axon returns to its normal resting polarized state after a few milliseconds. It is clear, however, that this behavior is consistent with observations of the action potential and is assumed to be the source of it. These observations are quantitatively described by a set of phenomenological differential equations called the *Hodgkin-Huxley relations* and will be discussed later.

The behavior of membranes is one of the least understood processes in modern biology and for that reason is one of the most actively studied. There are many research groups trying to unravel the nature of the active transport mechanism, the workings of the "gates" that open and close to permit or retard the passage of sodium ions, and the role of metabolism in nervous processes. So much has been learned about membranes that it is now possible to construct a number of different artificial membranes using various kinds of phospholipids. These artificial membranes have relatively large areas and can be studied under much more controlled conditions than one normally has with natural membranes.

We have seen how concentration differences can produce electrical potentials under circumstances where very little charge needs to be transferred. The ions exchanged during an action potential event occupy only a very thin layer of the fluid on either side of the membrane. Therefore the normal concentration differences that exist between the inside and outside of the cell can support a very large number of firings without any replenishment by the active pumping mechanism. It is clear that the active transport, a slow steady process supported by metabolism, provides the concentration difference from which the resting potential develops but is not the source of the potential per se. Changes from the resting potential occur when the permeability changes and the relative rate of flow of ions is changed.

Part 5: The Hodgkin-Huxley Equations

When a nerve membrane at its resting potential is suddenly depolarized, an action potential can result. Because the time course of the various currents in and out of the membrane cannot be easily sorted out by observing the pulse shape, a method was developed in 1949 by K. S. Cole to study the steady state behavior of membranes at various voltages. The idea is to apply a sudden step voltage to a membrane at resting potential and maintain the voltage by supplying whatever current is needed. The current demanded by the membrane to maintain the fixed voltage is measured. This *voltage clamp* apparatus is shown in Figure 6.13.

Electrode A is the input to a gain stabilized operational amplifier C whose other input is grounded. The output of C is therefore proportional to the membrane potential. The voltage V_0' derived from the battery is the desired membrane potential. The difference between V_0' and the output of C is amplified by the op amp D that delivers a current through electrode B to the inside of the axon. If there is any difference between the desired and actual membrane potential, op amp D generates a current whose sign is chosen to correct it. If there is no difference, the output of D is zero and the desired potential is therefore maintained. The membrane voltage is therefore clamped by a feedback control system similar to those we studied in Chapter 5.

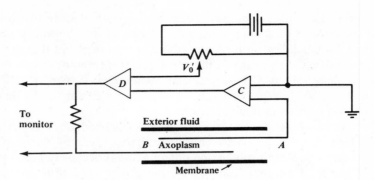

Figure 6.13 Circuit diagram of voltage clamp apparatus. Electrode A is the sensor whose signal is amplified by op amp C and compared with the set point by op amp D whose output is the effector in a feedback loop that maintains a constant membrane potential.

Experiments that shed light on the behavior of membranes are those with positive voltage steps V_0' applied to the B electrode. What currents flow through the membrane when its voltage is changed and maintained? To begin, the currents are not constant but change rapidly with time. In all cases, however, the first current delivered by op amp D is a pulse to depolarize the membrane. This current is positive for the (usual) positive step and lasts for only a fraction of a microsecond. It doesn't actually flow through the membrane but constitutes a displacement current across the membrane capacitance to discharge it from its resting potential V_0 to V_0'. This time constant of a fraction of a microsecond can be estimated from the resistance across an axon (through the axoplasm) and the capacitance. We find $R \cong pr/A = pr/2\pi xr = p/2\pi x$ where x is the length of the axon segment being studied. This is an approximation which is only valid if the radius of the electrode is not too much less than the radius of the axon. The correct expression is $R = (p/2\pi x)\ell n(r/r_e)$ where r_e is the electrode radius. Also, $C = A\epsilon/t = 2\pi xr\epsilon/t$ so that $RC = pr\epsilon/t \cong 0.01$ microsecond.

After the membrane has been depolarized, its properties change markedly. If the depolarization is great enough to exceed the threshold voltage, the membrane becomes very permeable to Na^+ ions and they come pouring in as a result of their lower concentration inside than outside. Even though they are positively charged, they can pass into the axon that has been raised to a positive potential because of the concentration difference. The increase in sodium permeability of the membrane occurs in about 1 msec and is followed by an equally rapid return to relative impermeability.

The depolarization of the membrane also affects its potassium permeability, but more slowly than the sodium permeability. Figure 6.14 shows that the

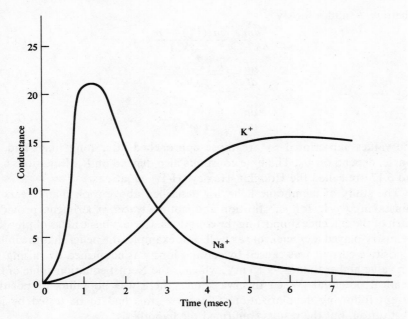

Figure 6.14 Plot of Na$^+$ and K$^+$ conductance versus time after the initiation of an action potential.

Na$^+$ permeability peaks and falls back to its resting level in a few milliseconds, whereas the K$^+$ permeability rises in many milliseconds and stays high (until the voltage step is removed). When the Na$^+$ permeability returns to its small resting value, the influx of Na$^+$ is halted, but the applied voltage continues to drive K$^+$ ions out of the axon through the permeable membrane as long as the voltage is maintained. When the voltage is shut off, the driving force is gone *and* the membrane becomes impermeable to K$^+$. The K$^+$ flow therefore stops.

In 1952 Hodgkin and Huxley derived a set of coupled differential equations that describe the time development of the Na$^+$ and K$^+$ conductance of membranes. They were chosen to fit the family of curves that would be generated if Figure 6.14 were plotted for many different values of V_0' instead of only one value. Of course they depend on V_0' in a very complicated way and are therefore parametrized in terms of three variables, m, h, and n, which depend exponentially on time with time constants τ_m, τ_h, and τ_n, respectively. These time constants are chosen to make the equations fit the data, which they do extremely well. The Na$^+$ and K$^+$ currents are given by

$$I_{Na} = g_{Na}m^3h(V_0' - V_{Na}), \tag{6.16a}$$

and

$$I_K = g_K n^4(V_0' - V_K), \tag{6.16b}$$

where m, h, and n satisfy

$$\frac{dm}{dt} = \frac{m_\infty(V_0') - m}{\tau_m(V_0')},$$ (6.17a)

$$\frac{dh}{dt} = \frac{h_\infty(V_0') - h}{\tau_h(V_0')},$$ (6.17b)

and

$$\frac{dn}{dt} = \frac{n_\infty(V_0') - n}{\tau_n(V_0')}.$$ (6.17c)

The values subscripted ∞ are those approached after long times and, of course, depend on V_0'. The time constants also depend on V_0'. Equations 6.16 and 6.17 are called the Hodgkin-Huxley (H-H) equations.

The study of membrane behavior described above took many years of painstaking work. Ion substitution and isotope tracer experiments provided much of the evidence supporting the equations. Certain basic ideas of physical chemistry played a crucial role as well. For example, the belief that the influx of positive current was caused by sodium ions was confirmed by raising V_0' to a value higher than $\cong +55$ mV, which is the Nernst equation value of the chemical potential. When this was done, there was no influx of positive current following depolarization. Of course, this had to be tested by ion substitution, but the results confirmed the hypothesis.

The H-H equations are only phenomenological descriptions of ion currents through membranes. They do not describe the fundamental physics of action potential propagation. Studies with artificial phospholipid bilayer membranes have shown that the ions are conducted through the membranes at particular locations called *channels* and that certain proteins are crucial to the actions of these channels. The problems are difficult to elucidate and the answers are complicated. Membrane research is currently one of the most active areas of biophysical research.

Part 6:   Communication Between Cells

When the action potential pulse reaches the end of the axon, knowledge of its arrival must be transmitted either to another nerve cell in a communication chain or to a muscle. In either case, the carrier usually is a *chemical transmitter* that diffuses across the small gap between the knob at the end of the axon and the special receptor site on the dendrite of the next cell. The chemical transmitter, usually *acetylcholine* or *norepinepherin*, is released in a short burst of a large number of molecules that either excite or inhibit a response from the receptor cell.

Figure 6.15 shows a schematic diagram of a synaptic junction between cells. The dendritic knobs contain small vesicles with a fixed amount of

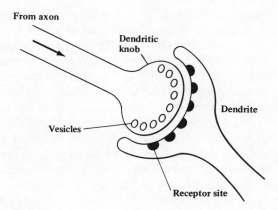

Figure 6.15 Sketch of typical synaptic junction showing dendritic kob and chemical transmitter vesicles.

chemical transmitter in each. In the normal resting state these are continuously released and the molecules diffuse across the gap to the receptor where they are bound to the proper site and then chemically destroyed. When the voltage from an action potential reaches the end of the nerve, however, the rate at which the vesicles are released increases by about 10^6 and a tremendous burst of them diffuses across the gap.

We normally think of the process of diffusion as being somewhat slow, and certainly unsuited as the primary mechanism for communication between cells in a system that is characterized by high-speed transmission. Diffusion takes place as a result of the random collisions between molecules and can therefore be modeled statistically. If we ask how far from its starting place a particular molecule can be expected to be found in a certain time, the probability distribution can be found from the random walk model. In this case, the distance varies as the square root of the time (see Appendix G). It takes nine times longer to diffuse three times further. On the other hand, short distances can be covered very rapidly by the random walk process, and it is for this reason that diffusion allows a nervous system to have communication at adequate speeds. Synaptic junctions are typically 100 to 200 Å wide, and transmitter molecules can diffuse across them in a small fraction of a millisecond.

When the nerve impulse is received at the end of the axon, a series of complicated processes results in the release of a large number of vesicles of transmitter molecules. The vesicle walls are destroyed, the molecules are released to diffuse across the gap, and then they are bound to specific receptor sites. In the case of acetylcholine transmission, the molecules are destroyed shortly after they have been bound in the dendrite by an acetylcholinesterase called *eserine*. As soon as the bound transmitter molecules are destroyed, new ones are available to replace them. The result of this bombardment by transmitters is a substantial chemical perturbation at the membrane of the dendrite or the muscle cell, and electrical changes (depolarization for example) occur there. The flood of transmitter molecules is rather quickly cleaned out, however, and the junction is prepared to transmit the next pulse.

There are many different chemical transmitters, and each one has its own particular esterase for rapid destruction of it. In addition, there are blocking agents for some transmitters. The chemical complexity of the system lends itself to a number of variations from the rather simple function of pure communication. For example, it is easy to imagine an *inhibitory synaptic junction*. In this case, we need a normal (excitatory) junction accompanied by a second input that would serve to inhibit communication. It could work by releasing a blocking agent that kept the transmitter from being released, by destroying the transmitter before it can diffuse across, or by interfering with the esterase so that the junction could only transmit pulses at a reduced rate. Many nerve poisons are active in the synaptic junctions. Curare and rattlesnake venom both inhibit the binding of the transmitter on the synapse whereas botulin toxin prevents release of the transmitter.

Chemical transmitters, blockers, and destroyers (esterases) form several different families, each of which may have several members. In some cases there can be interaction between different family members and even between different families, but in other cases the interactions are highly specific. In general, the whole problem is very complicated and only partially understood.

The chemical communication channel has certain desirable properties. It provides a mechanism for connections and branches. Because of its extreme specificity, it is relatively noise-free and immune to a number of rather large perturbations. It is sufficiently redundant that it can withstand extensive damage and still continue to function. It is certainly superior to many man-made systems.

Part 7: Muscles

Although the topic of muscle structure and muscular contraction could easily be the subject of a whole book, it is limited to a section of the nerve chapter here because of space limitations and the similarities between muscles and nerves. A muscle is made of a large number of fibers with diameters between 10 and 100 microns. These fibers are often as long as the entire muscle and appear to have striations of light and dark across them [see Figure 6.16(a)]. Each fiber is bounded by an excitable membrane whose selective permeability is responsible for a voltage drop of about 10 mV (negative inside) just as in the nerve axon. When the muscle receives an impulse from a nerve, the

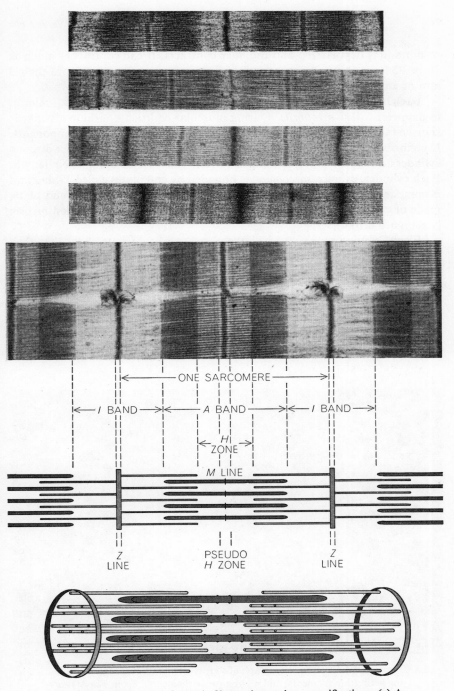

Figure 6.16 Diagrams of muscle fiber at increasing magnifications. (a) A myofibril with several sarcomeres. (b) The origin of the different bands of a sarcomere. (c) The anatomy of a single sarcomere composed of actin and myosin fibrils. From *The Mechanism of Muscular Contraction*, H. E. Huxley. Copyright © 1965 by Scientific American, Inc. All rights reserved.

membrane is depolarized and the fiber contracts. It can contract as much as 20 or 25% in about 200 msec and can generate a pressure of more than 3 atm or about 4×10^5 N/m².

Each muscle fiber is made of a number of parallel elements about 1 micron in diameter called *myofibrils*. During muscular contraction, these myofibrils shorten (and thicken slightly) as a result of a telescoping of their components. Myofibrils [see Figure 6.16(b)] are composed of a large number of a discrete cylinders about 2.5 microns long arranged end to end along their length. Each cylinder, called a *sarcomere*, is bounded by end plates called *Z-lines* and is composed of three arrays of thin filaments, one bound at each end plate, made of *actin*, and one array of thicker filaments in the middle called *myosin* [see Figure 6.16(c)].

Electron microscope views of these arrays of filaments show that they are arranged in a very regular, hexagonal close-packed array and that there is no structural connection between the myosin and actin arrays. Figure 6.17 shows the myosin (about 160 Å thick) and actin filament (about 60 Å thick) arrays as seen in cross section. The entire structure is kept from collapse by a large number of cross bridges between the myosin and actin filaments. Since it is clear that these cross-links cannot possibly be maintained during muscle contraction and relaxation, it is supposed that they disconnect and

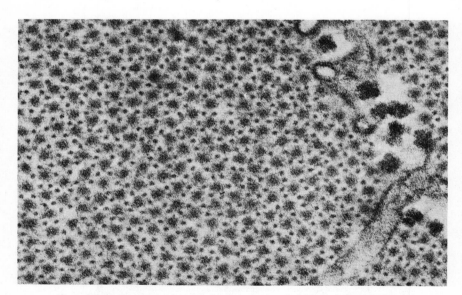

Figure 6.17 Electron micrograph of a cross section through muscle showing actin (thin filaments) and myosin (thick filaments). From *The Mechanism of Muscular Contraction*, H. E. Huxley. Copyright © 1965 by Scientific American, Inc. All rights reserved.

then reattach themselves at new positions. As the cross bridges "walk" along the filaments, they can actually generate the muscular force.

Each sarcomere is a muscle "unit". It can contract or relax under stimulation or lack of it from a nerve ending. It consumes energy when it contracts, and that energy is later released as heat. However, the production of movement by the cross-links between the actin and myosin filaments cannot proceed through the production of heat, or else the efficiency of muscle would be thermodynamically limited to some value below its observed value. It must be that the energy from splitting the *ATP molecule* (the basis of muscular energy) is converted directly to mechanical energy by the cross bridge during muscle contraction.

There are several regions of the sarcomere of different average density, and these appear as bands across the myofibril. They are given names [see Figure 6.16(b)] that are often referred to in the literature. In the early stages of muscle contraction, the H-zone, which is part of the A band, shortens until it almost disappears and then the A bands continues to shorten further as the actin fibers begin to overlap one another. It is important to remember that the band description only represents the appearance of the myofibrils and that the sliding-filament model of the sarcomeres represents the best knowledge we have about the mechanism of muscular contraction.

APPENDIX G
DIFFUSION

Diffusion is a transport process that results from the random collisions of molecules with one another. At first thought it seems that diffusion would not result in any net transport because there are so many random collisions that the net (average) displacement should be zero. On the other hand, the fact that the average or expected displacement is zero does not mean that the actual displacement is exactly zero in every case. If we consider the results of a large number of coin flips, N, the expected number of heads is $N/2$. But in any given sequence there is a rather large probability that there will not be $N/2$ heads; indeed it is possible that there will be zero heads.

In describing a random sequence which has events of interest which occur with probability p, we can write the probability that in N trials there will be n of these events as

$$P_n(N) = \frac{N!}{n!(N-n)!}p^n(1-p)^{N-n}. \tag{G.1}$$

For the case of coin flipping, $p = \frac{1}{2}$, and $P_n(N) = N!/[n!(N-n)!2^N]$, which has a maximum at $n = N/2$ but is certainly finite at other values of n. For example, we find $P_{10}(20) = 0.176, P_{11}(20) = 0.160$, and $P_{12}(20) = 0.120$; that is, the probability of getting 12 heads and 8 tails in 20 flips is 0.120, which is two-thirds of the probability of 10 each of heads and tails. Of course, the sum of the probabilities must always be unity:

$$\sum_{n=0}^{N} P_n(N) = 1. \tag{G.2}$$

We begin our discussion of diffusion by studying the random movements of a single molecule under the influence of collisions. We consider the case where the molecule moves in only one dimension, and each movement covers the same distance; that is, we allow equal steps of size $\pm s$ with the sign for each step determined randomly. We can easily calculate the displacement of a molecule after N such steps with n of them positive, $d_n(N)$, by computing the number of positive and negative steps and subtracting: $d_n(N) = ns - (N-n)s = (2n-N)s$. The net expected displacement $d(N)$ is found by taking the weighted average for positive and negative steps and subtracting:

$$d(N) = \underbrace{\sum_{n=0}^{N} nsP_n(N)}_{\substack{\text{positive} \\ \text{distance}}} - \underbrace{\left[Ns - \sum_{n=0}^{N} nsP_n(N)\right]}_{\text{negative distance}} = 2s\sum_{n=0}^{N} nP_n(N) - Ns. \tag{G.3}$$

We can evaluate the summation on the right side of Eq. G.3 by substituting the expression for $P_n(N)$ from Eq. G.1.

$$\sum_{n=0}^{N} nP_n(N) = \sum_{n=1}^{N} \frac{nN!}{n!(N-n)!} p^n(1-p)^{N-n}$$

$$= Np \sum_{n=1}^{N} \frac{(N-1)! \, p^{n-1}(1-p)^{N-1-(n-1)}}{(n-1)![N-1-(n-1)]!}. \tag{G.4}$$

The sum on the right side can be easily evaluated by substituting $k = n - 1$ and $K = N - 1$ and recognizing that the sum is equal to unity just as in Eq. G.2 so that we have $\sum nP_n(N) = Np$. We then find

$$d(N) = Ns(2p - 1) \tag{G.5}$$

which is Ns for $p = 1$, $-Ns$ for $p = 0$, and zero (as expected) for $p = \frac{1}{2}$.

We have shown that the random walk expectation value of $d(N)$ is zero and have argued that any given molecule is not likely to achieve a value of d that is exactly equal to the expectation value. We need to determine some measure of the spread of values of the displacement. As a matter of convention we use the average of the squares of the displacement. Since the squares are all positive, this average is not zero. Then σ_d, the measure of the spread of values of d, is given by

$$\sigma_d^2 = \langle [d_n(N) - d(N)]^2 \rangle = \langle [2n - N - N(2p - 1)]^2 \rangle s^2$$

$$= 4s^2 \langle (n - Np)^2 \rangle, \tag{G.6}$$

where the symbols $\langle \ \rangle$ mean average value of the quantity enclosed. We expand the squared bracket in Eq. G.6 and find that we have to evaluate the quantity $\langle n^2 \rangle$, which can be done using the following trick. We use $n^2 = n(n - 1) + n$ and observe that the quantity $\langle n(n - 1) \rangle$ can be found by evaluating $\sum n(n - 1)P_n(N)$ in a very similar way to that used in Eq. G.4 and G.5. The result is $\langle n(n - 1) \rangle = p^2 N(N - 1)$. Therefore

$$\langle n^2 \rangle = p^2 N(N - 1) + pN \tag{G.7}$$

which we emphasize is different from $\langle n \rangle^2 = N^2 p^2$. We can now evaluate Eq. G.6 for σ_d:

$$\sigma_d^2 = 4s^2(\langle n^2 \rangle - 2\langle nNp \rangle + \langle N^2 p^2 \rangle) = Np(1 - p)4s^2. \tag{G.8}$$

Notice that this spread in values is zero if $p = 0$ or $p = 1$ as we might expect. The spread is maximum when $p = \frac{1}{2}$ and is proportional to $\sqrt{N}$. This last feature, the $\sqrt{N}$ dependence, is extremely important for a number of reasons. We shall see it again in this appendix and also in Chapter 9.

For $p = \frac{1}{2}$, Eq. G.8 becomes

$$\sigma_d = s\sqrt{N} \tag{G.9}$$

from which we can derive an expression for the average velocity of a diffusing molecule. If τ is the time between collisions, the diffusion velocity v_d is

$$v_d = \frac{\sigma_d}{N\tau} = \frac{s}{\tau\sqrt{N}} \tag{G.10}$$

which is not very much smaller than the thermal velocity s/τ for small values of N; that is, diffusion is a rapid transport mechanism over fairly small distances. For example, if the distances traveled between collisions is $\frac{1}{100}$ of an atomic diameter, $s = 10^{-10}$ cm, then $\tau = s/(\text{thermal velocity}) = 10^{-10}/10^5 = 10^{-15}$ sec. In 1 millisecond there are 10^{12} collisions so that σ_d (1 msec) $= 10^{-4}$ cm (from Eq. G.9). Since a typical synaptic junction requires diffusion over about 100 Å $= 10^{-6}$ cm, we see that the communication between cells can occur in a small fraction of a millisecond.

We have discussed the one-dimensional random walk to show the general properties of the solutions. The arguments that produced the essential feature of these solutions, namely the $\sqrt{N}$ dependence, are readily extended to two and three dimensions. A more careful calculation will take into account the fact that the step sizes (distance traveled between collisions) are not all the same and that the time between collisions is not always the time. In fact, both s and τ have statistical distributions that can be folded into the calculation. It turns out that the result is described in terms of a single temporal parameter, the mean time between collisions, and a single spatial parameter, the mean distance between collisions. These are given by Eq. G.9 and G.10 with s and τ taking their mean values.

Diffusion occurs as a result of a concentration difference Δc between two separate regions of space and can be described macroscopically. Consider a region of space bounded by planes of area A separated by dz so that the volume they enclose is $A\,dz$. Suppose the concentration of molecules is c on one side of this region and $c + \Delta c$ on the other side. It is clear that the rate of molecules crossing the region is proportional to Δc because the rate of diffusion across the region in one direction is proportional to c and in the other direction proportional to $c + \Delta c$. The rate of molecules crossing the region dn/dt is also proportional to A and inversely proportional to dz. We write

$$\frac{dn}{dt} = DA\frac{dc}{dz} \tag{G.11}$$

where D is called the diffusion constant and depends on the detailed nature of the collisions responsible for the diffusion. We can differentiate Eq. G.11 and find

$$\frac{dc}{dt} = D\frac{d^2c}{dz^2} \tag{G.12}$$

where we have used the fact that $c = dn/A\,dz$. Equations G.11 and G.12 are alternative forms of very common equations describing a wide range of transport phenomena (heat flow, electrical current, etc.). When used to describe diffusion, they are called *Fick's law*.

Questions and Problems for Chapter 6

1. The resting voltage across each axon membrane in your leg is about 70 mV, and there are thousands of axons. Why isn't there a macroscopic voltage (few volts) across your leg?

2. Imagine an axis drawn perpendicular to the length of an axon through its center. Make a plot of the potential along this axis, starting from outside the axon, going through it and coming out again on the other side. Make another plot showing the concentration of K ions along the axis. What can you say about electrical neutrality? Estimate the concentration difference between the inside and the outside.

3. In order for an axon to propagate a pulse, the only requirement is an excitable membrane. It seems that the action potential should propagate in two directions simultaneously resulting in chaos to the nervous system. Why doesn't this happen?

4. Why does the reciprocity relation of Eq. 6.14 fail for long times? What is mathematically wrong with the equation at long times?

5. If the switch is closed in the accompanying circuit, the signal to light the bulb travels at the speed of light. Compare and contrast this signal with that carried by a nerve axon. Discuss your answer.

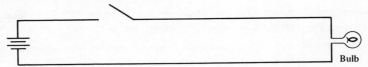

6. What is the difference between conduction of pulses by myelinated nerves and unmyelinated nerves?

7. Explain how a potential difference can arise across a membrane that has different permeabilities for the different ions in the solutions on either side of it.

8. What do we learn about nerves from the voltage clamp experiments? Why is this information important?

9. An action potential can be initiated in an axon by application of a positive voltage inside. The minimum value of this voltage required to initiate the action potential (threshold voltage) decreases if it is allowed to remain on for a longer time; that is, there is a tradeoff between pulse height and pulse duration. What are the limits to this tradeoff? What is the physical basis for the existence of this tradeoff?

10. What is an inhibitory synaptic junction? How is it different from an excitatory one? What purpose might it serve?

Bibliography

1. E. ACKERMAN, *Biophysical Science* (Prentice-Hall, Inc., Englewood Cliffs, N. J., 1962). (An excellent treatment of a variety of topics in biophysics.)

2. E. ACKERMAN, L. ELLIS, and L. WILLIAMS, *Biophysical Science* (Prentice-Hall, Inc., Englewood Cliffs, NJ 1979). (A new edition of the previous reference.)

3. CAROLYN COHEN, "The Protein Switch of Muscle Contraction," *Scientific American* (1976).

4. G. EHRENSTEIN, "Ion Channels in Nerve Membranes," *Physics Today*, **29**, 33 (October 1976). (Contains a good description of H-H equations.)

5. R. HOBBIE, *Int. Physics for Medicine and Biology* (John Wiley & Sons, New York 1978). (An excellent treatment of many important topics of great interest. Highly recommended.)

6. H. E. HUXLEY, "The Mechanism of Muscular Contraction," *Scientific American* (December 1965).

7. BERNARD KATZ, *Nerve, Muscle, and Synapse* (McGraw-Hill Book Co., New York, 1966). (The classic introduction to neurophysiology. Required reading for anyone with a special interest in this field.)

8. G. MATSUMOTO and I. TASAKI, *Biophysics J.* **20**, 1 (1977).

9. T. RUCH and J. FULTON, *Medical Physiology and Biophysics* (W. B. Saunders, Co., Philadelphia, Pa., 1960). (A standard physiology text.)

10. ROBERT SCHMIDT, ed., *Fundamentals of Neurophysiology* (Springer-Verlag, New York, 1975). (A careful, clear, easy-to-read introduction to the topic.)

11. A. SCOTT, *Rev. Mod. Phys.* **47**, 487 (1975).

12. E. SELKURT, ed., *Physiology* (Little, Brown and Co., Boston, Mass., 1971). (A standard physiology text.)

7

SOUND AND HEARING

Part 1 : Vibrations and Waves

The sensation of sound results from our ears detecting vibrations in the air that are propagated as waves from the sound source to our ears. We therefore begin the study of sound and hearing with a description of vibrations and waves. The study of vibrations will be of much broader use than just acoustics. The ideas presented in this section will be used again in Chapter 9 when we study molecular motion.

The simplest imaginable vibrating system, called a *harmonic oscillator*, consists of a point mass mounted on one end of an ideal massless spring. The force exerted by the spring on the mass is proportional to the amount x by which the end of the spring is moved from its equilibrium position (x is not the length of the spring!) and is in the opposite direction to the displacement. We write this force as $F = -kx$. If the spring force is the only one acting on the mass, the definition of force, $F = dp/dt$, gives us the equation of motion $ma + kx = 0$ since $p = mv$, m is constant, and $a = dv/dt$. Since the acceleration a is in the x direction, we call it $\ddot{x} = d^2x/dt^2$ and we have

$$m\ddot{x} + kx = 0. \tag{7.1}$$

There are many ways to write the solutions to Eq. 7.1. If we let

$$x = A \cos(\omega t + \phi) \tag{7.2}$$

we find that A and ϕ must be determined by the initial conditions of the problem and that $\omega = \sqrt{k/m}$. For example, if the mass is released from rest at $t = 0$ after being displaced by 5 cm, $A = 5$ cm and $\phi = 0$. In order to facilitate the solution of some other problems we shall discuss later, we choose to write the solutions in the form $x = Ae^{i(\omega t+\phi)}$ or equivalently $x = A \exp[i(\omega t + \phi)]$ where we must always remember to take the real part of any final result we calculate (recall that $e^{i\theta} = \cos\theta + i\sin\theta$).

Let us now consider the case of a harmonic oscillator that is characterized by some damping or friction. We consider only the simple case where the friction force is proportional to the velocity $\dot{x} = dx/dt$ and in the opposite direction. Then the friction force is $-\beta\dot{x}$ (where β is some constant) and the equation of motion is

$$m\ddot{x} + \beta\dot{x} + kx = 0. \tag{7.3}$$

We again try $x = A \exp[i(\omega t + \phi)]$, which we substitute into Eq. 7.3 and find

$$-m\omega^2 + i\beta\omega + k = 0. \tag{7.4}$$

This is solved for ω using the standard quadratic formula. It is convenient to define $\omega_0 = \sqrt{k/m}$ and $\gamma = \beta/2m$ and then the solutions for ω are

$$\omega = i\gamma \pm \sqrt{\omega_0^2 - \gamma^2}. \tag{7.5}$$

We notice that in the case of zero friction ($\gamma = 0$) Eq. 7.5 reduces to the first example as it should, but the notion of a complex angular frequency for the nonzero friction case is a little confusing. It is premature, however, to simply take the real part of Eq. 7.5. We must substitute this value of ω back into our expression for x and then we find

$$x = A \exp[i(i\gamma t)] \exp(i\sqrt{\omega_0^2 - \gamma^2})t \exp(i\phi). \tag{7.6}$$

The first exponential factor becomes $\exp(-\gamma t)$ and represents simply an exponential damping of the oscillations. The second exponential factor is purely oscillatory, but at a slightly lower frequency than the nonfriction value of $\sqrt{k/m}$.

The third factor (phase) and the amplitude are again determined by initial conditions. The solutions of the damped oscillator have led us to an exponentially decaying oscillation at a frequency slightly lower than ω_0. The length of time it takes for the oscillator to decay to $1/e$ of its original amplitude is called the *decay time*, and the number of cycles in one decay time is called the *quality factor*, Q. A plot of the position versus time of a damped oscillator is shown in Figure 7.1; the dotted lines are called the *envelope* of the oscillation. We have seen that retaining the complex frequency leads to a physically sensible solution.

The next oscillatory motion we consider is the case where the mass is driven by some oscillating force. We describe the oscillating force in terms of an exponential in order to make it compatible with the solution. We write

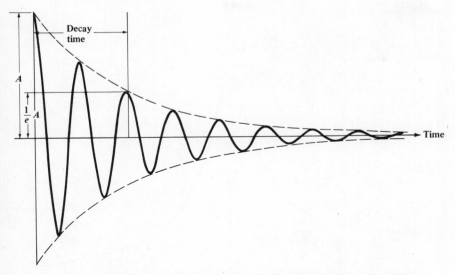

Figure 7.1 Plot of damped oscillation versus time showing exponential envelope and decay time.

the equation of motion

$$m\ddot{x} + \beta\dot{x} + kx = F_0 \exp{(i\omega t)} \tag{7.7}$$

where ω is the angular frequency of the driving force. We assume the mass eventually oscillates at the frequency of the driving force and again try a solution $x = A \exp{[i(\omega t + \phi)]}$, which leads to

$$(\omega_0^2 - \omega^2)A + 2i\omega\gamma A = \left(\frac{F_0}{m}\right) \exp{(-i\phi)}. \tag{7.8}$$

Again $\omega_0 = \sqrt{k/m}$ and $\beta = 2m\gamma$. Since the angular frequency ω is known (it's the driving frequency) the unknowns are the phase ϕ and the amplitude A (which are no longer simply a matter of the initial conditions). We solve for these by separating the real and imaginary parts of Eq. 7.8. We find

$$\tan\phi = \frac{2\gamma\omega}{(\omega^2 - \omega_0^2)} \tag{7.9a}$$

and

$$A = \frac{F_0/m}{\sqrt{(\omega_0^2 - \omega^2)^2 + 4\gamma^2\omega^2}}. \tag{7.9b}$$

These are very interesting solutions and should be studied carefully. We notice that the amplitude depends on the driving frequency and that it has a maximum near $\omega = \omega_0$ as shown in Figure 7.2(a) (called a *resonance curve*). The points at which the height is one-half the maximum value are separated by an amount proportional to γ and their frequency separation is called the

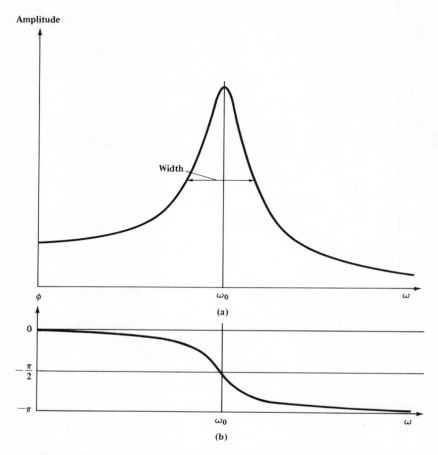

Figure 7.2 (a) The amplitude versus frequency of a driven, damped oscillator. Note that it is not zero at zero driving frequency but is zero at very high frequencies. The width is determined by the damping coefficient. (b) The relative phase of the oscillator versus driving frequency. It changes from zero to 180° out of phase in a frequency range comparable to the width of the amplitude curve.

width of the resonance. When the width is small compared to the natural frequency ω_0, we say the system has a high Q. We note that the amplitude would become infinite if there were no friction. We call ω_0 the *natural* or *resonant frequency* of the system because it is the vibration frequency of the undamped free oscillator. We note that for small values of ω the amplitude is fairly constant and independent of ω but that as ω increases toward ω_0, the dependence becomes stronger. The approach to a stronger amplitude dependence on driving frequency depends on γ and, in fact, determines the width of the peak. Also note that for small values of the driving frequency the phase is near zero. As the driving frequency approaches resonance, tan ϕ becomes negative and infinite and ϕ is, therefore, $-\pi/2$. At high frequencies the sign of the phase is such that the oscillations are 180° out of phase with the driving force as shown in Figure 7.2(b).

The simple resonant system we have studied above is extremely important to the study of vibrating systems. Any rigid structure has some elastic behavior that can be described in the terms of an oscillator for small vibration amplitudes. Furthermore, there is at least one frequency where the oscillation amplitude is resonantly enhanced by the mechanical properties of the structure. In most cases we seek structures that have very broad resonance curves (musical instruments, buildings, concert halls) for obvious reasons.

Most of the systems we normally encounter cannot be described as a point mass on an ideal spring. Usually the mass is distributed over some volume of space, and there are many forces between the small elements of the mass that make up the whole. In order to form a physical picture of such a more realistic system, we now consider the case of two masses bound together by a spring and fixed to some laboratory framework by two other springs [see Figure 7.3(a)]. The masses are constrained to move along the

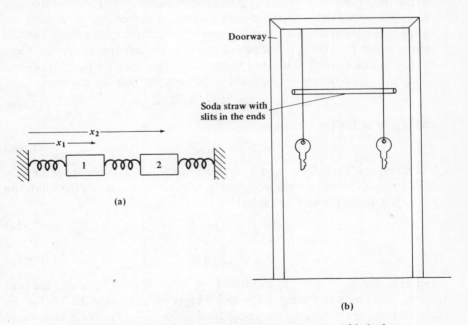

Figure 7.3 (a) Coupled oscillators connected by springs on a frictionless table or air track. (b) An arrangement of more readily available items that illustrate the essential features of coupled oscillators.

same straight line, and we study cases where the amplitude is small enough so that they don't hit each other. We call this a system of *coupled oscillators.*

When such a system is displaced from equilibrium and released, the motion is very complicated. It does not exhibit the simple oscillatory characteristic of the single oscillator but instead seems to execute a complicated series of irregular movements in which one mass frequently appears to stop completely for some time. This irregular motion does not seem to be describable in simple terms.

A little careful study will soon show that this system can be made to behave simply. For example, if both masses are displaced to one side keeping the distance between them unchanged, then there will be no force on either mass from the middle spring. Each of them will, therefore, oscillate as if they were driven only by the outer springs connected to them, and the distance between them will always be fixed and constant. The motion is easily interpreted as a simple oscillation. Similarly, if the masses are equally displaced in opposite directions and again released from rest, they will also oscillate in a simple manner. These particularly simple vibrations are called *normal modes*, and it is not difficult to find a mathematical description of them.

We label the distance from the left fixture to each mass x_1 and x_2 and denote the equilibrium values of the coordinates by x_1^0 and x_2^0. For the sake of mathematical simplicity (but with no loss of generality) we assume the springs are not stretched when the masses are at x_1^0 and x_2^0. We study how the masses move by writing equations of motion. The left spring exerts a force $k(x_1^0 - x_1)$ on the left mass, and the center spring exerts a force $k[x_2 - x_2^0 - (x_1 - x_1^0)]$ on that mass. Neglecting friction, we have the equation

$$m\ddot{x}_1 = k(2x_1^0 - 2x_1 + x_2 - x_2^0), \qquad (7.10a)$$

and similarly for the other mass

$$m\ddot{x}_2 = k(2x_2^0 - 2x_2 + x_1 - x_1^0). \qquad (7.10b)$$

These two equations for x_1 and x_2 in terms of their derivatives are coupled in such a way as to make them inseparable. We can, however, do the following trick. We define the new variables

$$Y_1 = (x_1 - x_1^0) - (x_2 - x_2^0), \qquad (7.11a)$$

and

$$Y_2 = (x_1 - x_1^0) + (x_2 - x_2^0), \qquad (7.11b)$$

and note that $\ddot{Y}_1 = \ddot{x}_1 - \ddot{x}_2$ and that $\ddot{Y}_2 = \ddot{x}_1 + \ddot{x}_2$ since x_1^0 and x_2^0 are constants. Then we add the Eq. 7.10a and b together, substitute the Y's for the x's, and find the astonishingly simple result $m\ddot{Y}_2 = -kY_2$. We also subtract Eq. 7.10a from b, again substitute Y's for the x's, and find $m\ddot{Y}_1 = -3kY_1$, which is also extremely simple. The solutions of these equations for the Y's are simple and well-known to us.

Observe that Y_1 is simply a measure of the separation between the two masses and that Y_2 is a measure of their net displacement from equilibrium. If the masses are displaced in opposite directions and released so that they oscillate in opposition, the motion is described by an oscillatory behavior of the variable Y_1. Similarly, if they are displaced in the same direction and released so that they oscillate in unison (as if the coupling spring were made into a rigid rod), the motion is described by an oscillatory behavior of Y_2. We, therefore, interpret the Y's as the variables that describe the normal modes of the system. We notice that the normal modes occur at different frequencies: $\omega_1 = \sqrt{3k/m}$ and $\omega_2 = \sqrt{k/m}$.

The reader is urged to study another system of coupled oscillators. Two pendula, made by hanging weights (keys) on one-meter lengths of thread in a doorway can be coupled with a light, rigid rod (a soda straw is ideal). In Figure 7.3(b), an illustration of the proposed experiment shows the approximate proportions. Try releasing one of the pendula after displacing it a few centimeters from its normal equilibrium position and watch the motion. Then try to observe the normal modes of oscillation (similar to those of the spring coupled masses). Finally, move the coupling bar to a different height and repeat the experiments.

We could readily extend the study of the coupled oscillator problem to include the effects of friction and then consider the case where there is also a driving force. We would find that there are two resonant frequencies (near those of the normal modes) and that the resonance curve has two peaks. If the Q of the system is low enough (large β) the two peaks would blend into one broad one.

The normal modes of systems of many masses that can vibrate in more than one dimension are extremely complicated and not easy to find. Under certain conditions, however, the normal modes of some complicated systems are particularly simple and easy to describe because they constitute what we call *wave motion*. It is important to realize that wave motion together with its diffraction, refraction, reflection, and interference is simply a superposition of a small number of normal modes of a very complex vibrating system. The propagation of sound waves falls into this category.

Sound waves are produced by a variety of complicated vibrating systems. The sound board of a piano and the human voice box are only two examples of a large class of mechanically complicated arrangements whose vibrations can be stimulated or controlled to produce desired (and undesired) vibrations. The common feature of many sound sources is a structure with a very large number of closely spaced normal modes that can be resonantly excited by driving forces over a large range of frequencies. The vibrations are established in the air and propagate as sound waves. These sound waves can cause vibrations in our ears which are transduced to electrical signals that are sent to the brain and interpreted as the sensation of sound.

Part 2: Sound

As sound waves are propagated through the air (or other medium), there are compressions and rarefactions distributed through the space. Regions of compression (maximum pressure) are called *crests* and regions of rarefaction (minimum pressure) are called *troughs*. The separation between these crests or troughs is called the *wavelength*, and the number of times the pressure oscillates in one second is called *frequency*. We also recall that the product of f, the wave frequency, and λ, the wavelength is simply the velocity v at which the wave propagates.

The density of air molecules and the fluctuations caused by sound waves can be found from the kinetic theory of gases. We know that at room temperature and atmospheric pressure there are 6×10^{23} molecules in 22.4 liters of air. Consequently there are about 2.6×10^{19} molecules in 1 cm^3, and the average spacing between them is therefore about 3.3×10^{-7} cm. This is about 30 times larger than their diameter. When a sound wave propagates through the air, the density fluctuations have an amplitude that depends on the sound intensity. For normal human speech the pressure fluctuation amplitude is about 1 dyne/cm^2, which means that the density of air fluctuates by about 1 part in 10^6 (atmospheric pressure is about 10^6 dyne/cm^2). This means that the average spacing between the molecules only changes by about 3 parts in 10^7. To appreciate the small size of this fluctuation, note that if you stopped breathing for this fraction of your life, it would only take about one minute, an average of one second per year. It is astounding that our hearing mechanism can easily detect such tiny variations. At its limit, the hearing mechanism can detect variations 10,000 times smaller than this.

The mechanical vibrations whose frequency and amplitude enable them to be detectable by the human ear are called *sound waves*. Generally the frequency of these waves falls between 20 and 20,000 Hz and since they are normally heard in air where the speed of sound is about 320 m/sec, the wavelengths range from 16 m to 16 mm. We normally interpret changes in frequency as changes in pitch and find that most music and speech falls within the range from 100 to 3000 Hz.

We commonly describe the intensity of sound waves in terms of the pressure difference between a crest (maximum) and a trough (minimum). We call half the difference between these extremes the *amplitude* of the wave, but we must be careful to distinguish it from the amplitude of a simple oscillation. The lowest detectable amplitude of sound waves corresponds to a pressure fluctuation of about 2×10^{-4} dyne/cm^2, which is about 10^{-10} atm. This lowest audible sound defines the threshold of human hearing and can only be detected when the frequency is near 3000 Hz. At all other frequencies, the threshold is higher. Figure 7.4 shows the threshold of hearing versus frequency.

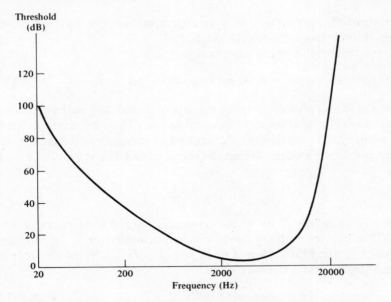

Figure 7.4 The audibility threshold curve determined from years of laboratory measurements.

Sound waves carry energy from the source in the form of compression and rarefaction of air (or other transmitting medium). The work done by this compression is the energy of the wave. We calculate the energy per second, or power, which is incident on a unit area. Power/area is called *intensity*.

The force F of the sound wave on an element of air volume is

$$F = pA \sin \omega t \qquad (7.12)$$

where p is the pressure amplitude of the sound wave and A is the area perpendicular to the direction of sound propagation. The power is simply the product of the force on an element of volume of air multiplied by its velocity. In order to calculate the velocity we use the expression

$$x = x_0 \sin \omega t \qquad (7.13)$$

for the time-dependent position of an element of air. The derivative of Eq. 7.13 is

$$\frac{dx}{dt} = v = \omega x_0 \cos \omega t. \qquad (7.14)$$

We estimate the magnitude of x_0 as the fraction of a quarter wavelength equal to the ratio of the wave pressure amplitude to atmospheric pressure p_A; that is, if an element of air were pushed a full quarter wavelength in each direction during the passage of a sound wave, the amplitude of the wave would be one atmosphere. Of course this is only an approximation, but it gives a physical insight into the more rigorous calculation.

The intensity of a sound wave is then

$$I = pAv \sin(\omega t)/A = p\omega x_0 \sin \omega t \cos \omega t = p\omega \frac{p}{p_A} \frac{\lambda}{4\pi} = \frac{p^2}{p_A} \frac{v}{2} \qquad (7.15)$$

where λ is the wavelength, v is the velocity of sound, and we have taken the time average of $\cos \omega t \sin \omega t$ to be equal to $1/\pi$. (This time average is taken over a half cycle to account for the fact that an energy detector absorbs completely and does no work back on the sound wave.) The velocity of sound in air is

$$v = \sqrt{\frac{1.4 p_A}{\rho}} \qquad (7.16)$$

where ρ is the density of air. At room temperature and atmospheric pressure, $v = 325$ m/sec. We multiply numerator and denominator of Eq. 7.15 by v, replace v^2 by using Eq. 7.16, and find $I = (\frac{7}{10})(p^2/\rho v)$, which differs only by a constant from a more rigorous result

$$I = \frac{p^2}{\rho v}. \qquad (7.17)$$

Note that the energy carried by a sound wave depends on the square of the pressure amplitude.

If the amplitude of a sound is changed, we interpret it as a change in loudness. The loudest sound we can stand without pain corresponds to a pressure amplitude about 1,000,000 times greater than threshold. For convenience we define the loudness scale of sound in terms of decibels (dB), named after Alexander Graham Bell. The dB scale is logarithmic, and if we call the amplitude of a threshold sound $p_0 = 2 \times 10^{-4}$ dyne/cm², then the loudness L of a sound in decibels is $L = 20 \log(p/p_0)$ where p is the pressure amplitude of the sound to be specified. Sometimes it is convenient to deal with sound levels in terms of the energy flux per unit area. The energy depends on the square of the pressure, so the intensity of a sound in decibels is $I = 10 \log(W/W_0)$ where W is the intensity of the sound to be specified in watts/per square centimeter and W_0 is approximately 10^{-16} watt/cm² (see Eq. 7.17). (W_0 corresponds approximately to a p_0 of 2×10^{-4} dyne/cm² in dry air at room temperature.) Note that the dimensions of W_0 are of mixed systems, mks and cgs, but this is common usage. Typical quiet situations correspond to a background noise level of about 20 dB, normal conversation occurs at a level of about 50 dB, the wind noise in a car traveling at 60 mph is about 75 dB, and the sound of an aircraft taking off overhead is about 110 dB. The threshold of pain is about 130 dB and permanent damage can be done to the auditory mechanisms by even a short exposure to sound at a level of 150 dB; 150 dB corresponds to vibrations whose amplitude is less than $\frac{1}{100}$ of atmospheric pressure.

Most of the sounds we hear are not simply describable as an oscillation having a well-defined amplitude at a single frequency. Most musical sounds are combinations or superpositions of a number of pure sine waves. Each of these component pure tones can be specified by its frequency and amplitude, but a description of the combination of them that constitutes the sound requires a list of each tone and its amplitude. This list is often presented in the form of a graph of amplitude versus frequency and is called a *spectrum*. The spectrum of a typical musical tone is shown in Figure 7.5, along with the waveform. Naturally, these two graphs are related, because each one is a

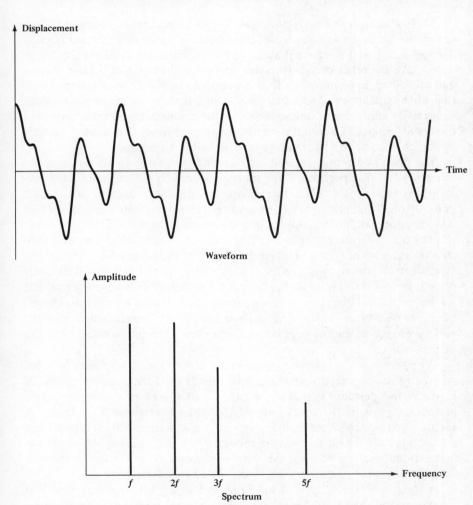

Figure 7.5 The waveform (a) and spectrum (b) of a relatively simple musical tone.

description of the other. The formal mathematical relation between them is called a *Fourier transform*. Musical tones are characterized by the fact that the description by spectrum is valid over relatively long periods of time during which the spectrum is constant. If the spectrum changes substantially during time intervals of the order of milliseconds, we say the sound is noise. If it is constant over time intervals of hundreds of milliseconds, we call it a tone. This means that a sound we call musical is one that has a repetitive waveform. As an example, we note that the spectrum of the coupled oscillators of Part 1 consists of two vertical lines, one at $\omega = \sqrt{3k/m}$ and the other at $\omega \sqrt{k/m}$.

Musical sounds can be further classified as consonant or dissonant. Consonant or harmonious sounds are usually those composed of a number of discrete tones, and the tone that is the lowest frequency is called the *fundamental*. All the other components are integral multiples of this fundamental and are called *harmonics*. The fourth harmonic is of a frequency four times that of the fundamental, etc. Needless to say, there is room for tremendous subjectivity and cultural influence in the distinction between consonant and dissonant sounds. If a sound is composed of pure tones that are not integral multiples of some lowest tone, we tend to call it dissonant.

The pitch of a musical sound is determined primarily by its fundamental frequency but the spectrum of the tones produced by various musical instruments is what differentiates them from one another. One can distinguish notes of the same pitch played on a variety of instruments because each of them produces different amounts of the various harmonics. Stringed instruments usually produce a sound rich in harmonics that often contains substantial amounts of power in all harmonics up to the twenty-fifth or thirtieth. Wind instruments usually produce tones with only the first few harmonics present in any substantial amounts, and brass instruments have spectra that fall between these extremes. It is the harmonic content of a sound that determines its quality or *timbre*. The same is true of the human voice, and the various characteristics of popular vocalists stem from the harmonic content of their voices.

When we listen to a pure tone or sine wave, the sound we hear is called pure or gentle or sweet. As various amounts of the harmonics are added to it, the sound develops character or fullness. If there are too many higher harmonics present, the tonal character might become harsh or sharp. A square wave, which sounds rather like a buzzer of a doorbell, is composed of a fundamental and all the odd harmonics of it, each having an amplitude proportional to $1/n$ where n is the order of the harmonic. Figure 7.6 shows a variety of musical spectra and their associated waveforms.

A fundamental and its first harmonic are separated by a factor of 2 in frequency, and this interval is called an *octave*. Since each octave represents a factor of 2 in frequency, the seven octaves of the piano correspond to an

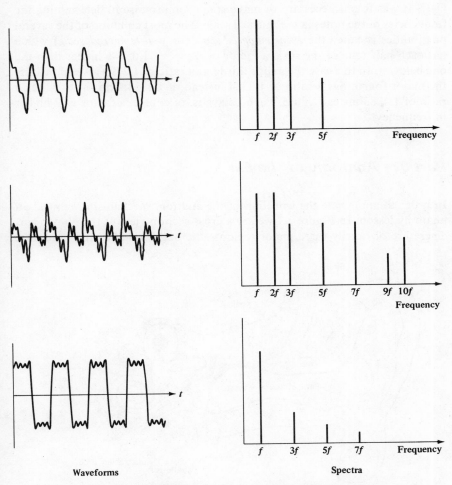

Waveforms   **Spectra**

Figure 7.6 The waveforms and spectra of several musical tones.

increase of 128 times the frequency from the lowest note to the highest. (See Appendix H for a discussion of musical notes.) When two notes are separated by a factor of $\frac{3}{2}$ in frequency, we call the interval a *musical fifth*. The ancients knew that two tones whose frequencies were in the ratio of small whole numbers were harmonious. The frequency ratio of two tones an octave apart is 2 and for two tones a fifth apart is $\frac{3}{2}$. Similarly, the frequency ratio of the tones in a major fourth is $\frac{4}{3}$ and in a major third is $\frac{5}{4}$. Problems arise, however, when we try to construct a system of musical intervals based on these simple ratios. A progression through 12 fifths (CGDAE, etc.) should bring one back to the original note seven octaves higher, but 2^7 is 128, not 129.7, which is

$(\frac{3}{2})^{12}$. It was found necessary to make some compromise in determining the frequencies of the notes in the musical scale. The most common of the several possibilities is called the *even-tempered scale* (or *well-tempered scale*) which has each half tone of the scale a factor of $(2)^{1/12} = 1.05946$ higher than the one below it. In this case the major third has a ratio of 1.2599 instead of 1.25, the major fourth has a ratio of 1.3384 instead of 1.3333, and the fifth has a ratio of 1.4983 instead of 1.5. Each octave is, of course, a factor of 2 higher in frequency.

Part 3 : Anatomy of the Ear

In order to appreciate the workings of the auditory mechanism, we need an anatomy lesson. In Figure 7.7 we see a cross-section view of the human ear. There are obviously three major regions: the outer, middle, and inner ear.

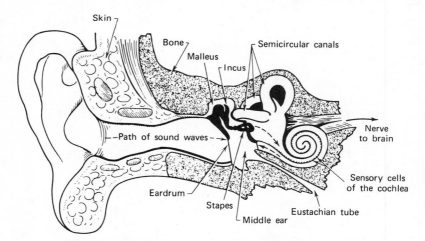

Figure 7.7 Gross anatomy of the ear as seen from a section taken from the front of the head. Major structures are labled. From Geldard, *The Human Senses*, 2nd ed. John Wiley and Sons, NY, 1972.

The outer ear consists of the *pinna*, which is the gristly, flat, and convoluted structure outside the head, along with the *auditory canal*, which is about 6 mm in diameter and about 3 cm long. It is terminated on the inside by the *eardrum*. We note that since it is an air column open at one end, it is resonant for waves whose length is four times its own length or about 12 cm. Since the speed of sound is about 32,000 cm/sec in air, the resonant frequency is about 2700 Hz. It is significant that the maximum sensitivity of human hearing occurs near this frequency. (See Figure 7.4.)

It is instructive to calculate the sensitivity of the ear. At 3000 Hz, the minimum detectable sound corresponds to an oscillation amplitude of about $x_0 = (p/p_A)(\lambda/4) = 10^{-9}$ cm! This is an extraordinarily small distance, about one-tenth the diameter of an atom. The pressure required to produce this vibration amplitude is about $p = 2 \times 10^{-4}$ dyne/cm². Since the area of the eardrum is only about $A = \frac{1}{2}$ cm², the total power incident on the ear is $(W_0/2) = 5 \times 10^{-17}$ watt. This is an incredibly small amount of power—it is the equivalent of only 200 optical photons per second.

The middle ear contains a chain of three tiny bones that serve to transmit vibrations from the eardrum to the actual organ of hearing, the *cochlea*. The mechanical advantage of the leverage system of the transmission bones is such that the amplitude of vibration at the *oval window* entrance to the cochlea is reduced in exchange for an increase in the pressure; that is, the eardrum-*ossicle* system serves as an acoustical impedance matching system to convert a pressure fluctuation into a mechanical motion and back to a pressure fluctuation of larger amplitude, but working on a smaller area. Naturally energy is conserved in this system as it is in any transformer.

Another important feature of the structure of the middle ear stems from the muscles that support and connect the ossicles. Under certain circumstances, these muscles can be contracted and thereby reduce the efficiency of transmission of vibration from the eardrum to the cochlea. This occurs when the ear is subject to very loud sounds and serves to protect the delicate inner ear from the trauma associated with large vibrations. Since it is a reflex action, it cannot act fast enough to protect us from sudden loud sounds such as a thunder clap, but it can ameliorate the discomfort from a slowly arising sound such as the arrival of a train in a subway station. In this regard, this *acoustic reflex* is very similar to the pupillary reflex of the eye.

The middle ear is vented to atmospheric pressure via the *Eustachian tube* that opens to the throat. Sometimes this tube becomes plugged with mucous associated with a cold or allergy and the middle ear cannot properly accomodate to pressure changes. The result of this situation is extreme discomfort that many people have experienced in air travel or a sudden change in altitude.

The primary auditory mechanism is the cochlea in the inner ear (see Figure 7.7). It is best described as a small tapered tube, divided along its length by the *basilar membrane*, and coiled up much like a snail. Its diameter is about 3 mm so its volume is about 15 microliters—about that of a couple of drops of water. The cochlea is solely responsible for the transduction of sound into nerve impulses and for much of our tonal discrimination capability. It is, therefore, the heart of our auditory system.

In the schematic cross section through the cochlea in Figure 7.8, we see that the basilar membrane has several structures built onto it. In particular, the *organ of Corti*, which houses the *hair cells*, and the *tectorial membrane*,

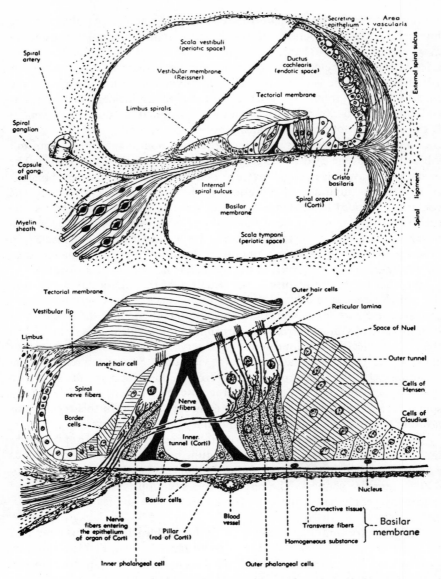

Figure 7.8 Cross sections through the cochlear canal. From Ruch and Fulton, *Medical Physiology and Biophysics*, 18th ed. W. B. Saunders Co., Philadelphia, 1960.

which is separately supported, are arranged so that a small movement of the basilar membrane results in the tectorial membrane rubbing across the ends of the hair cells. The long hairs provide leverage so that a relatively small motion of the basilar membrane or the cochlear fluid can be amplified enough to cause an excitation of the hair cells. Although the nature of this excitation is clearly some kind of depolarization of the hair cell membranes (which are more strongly polarized than the usual 70 mV), the mechanism of the process has not yet been elucidated.

We can see that the auditory function of the outer ear is to channel the sound waves to the middle ear. The middle ear transforms the low pressure oscillations to a higher pressure more suitable for detection and then transmits the oscillations into the cochlea. In the inner ear the oscillations are transduced from mechanical motion into electrical signals that are then transmitted to the brain.

Part 4 : Theories of Hearing

We have seen how the vibration energy of a sound wave is physically conducted to the nerve cells that transduce the sound to a nerve signal. We now ask how the signal is transmitted to the brain so it can interpret them as meaningful sounds. It is necessary to code sounds for transmission in some way since nerves can conduct information only as a series of spikes and not as a time varying waveform. The simplest question to ask is how the information about the frequency and amplitude of a pure sine wave is coded and transmitted, for once this is understood, any tone can be constructed by Fourier synthesis.

The most simple and naive model we can imagine is that of the basilar membrane oscillating up and down with each cycle of the sound wave. Its oscillatory motion is at the same frequency as that of the incident sound wave, and its amplitude varies monotonically with incident amplitude (not necessarily linearly). Its motion results in a stimulus to the hair cells that transmit a depolarization pulse along their axons at the peak (or valley) of each oscillation. As the amplitude of the incident sound increases, each axon may carry more pulses at the peak (or valley), or it may be that a larger fraction of the total number of axons participates in transmitting the sound. As the frequency increases, each axon must signal the more rapid arrival of each cycle by pulsing at a higher rate. In this way, all the information about an arbitrary sine wave can be coded and transmitted to the brain.

There are several problems associated with this theory of hearing (called *telephone theory* for obvious reasons). One of these problems stems from the inability of nerve axons to respond at a frequency of higher than about 1000 pulses per second simply because the refractory time between pulses is at least 1 msec. It would be impossible for an axon to transmit frequencies in the upper range of the auditory spectrum, and in the lower range it might be easy to confuse an intensity change with a frequency change. For these and other reasons, the telephone theory cannot be accepted as a complete theory of hearing.

Telephone theory can be modified somewhat to overcome these objections partially. The motivation for trying to save it stems from its simplicity as well as from data gathered in other experiments that indicate that there must be some truth in its ideas. The most important of these experiments is described by Gerald Oster in the October, 1973 issue of *Scientific American*. It is well-known that when two sine waves of different frequencies are combined, there results a signal at the sum and difference frequencies (see section 9.6, especially Eq. 9.32). Oster reports the results of stimulating one ear with one frequency and the other ear with another frequency. The subject can hear the difference frequency, which indicates that the original frequencies, and not some reduced coding of them, must be present in a brain area where the information from both ears is brought together. These *binaural beats* occur only in the lower-frequency ranges of audible sound. Another motivation for trying to save telephone theory is the simple fact that there is no other complete theory of hearing to replace it.

The *volley theory* of hearing is a modification of telephone theory that is based on the idea hinted at previously, namely, that not all axons may transmit a pulse at each peak of a waveform. The proposal is that each axon may transmit a pulse at some of the peaks, but not always. The summation of all the axon outputs, as shown in Figure 7.9, may constitute a very good reproduction of a high-frequency sine wave. The volley theory makes more efficient use of the axons in the auditory nerve and is more tolerant of errors and defects in the transmission line. Volley theory extends the useful range of the ideas of telephone theory to higher frequencies, but it is still not a practical theory for how we hear high frequencies.

In 1960 Georg von Bekesy published the results of decades of work on the human auditory system. His theory of hearing is based on the idea that the basilar membrane's structure permits its vibration amplitude to vary over its length and to have a maximum amplitude at some point along the membrane. The frequencies and amplitudes of the normal modes of vibration of this structure depend on the thickness and stiffness of the membrane at different points. For each frequency, there is one section of the membrane that is closer to resonance than any other section, and it is here that the maximum vibration amplitude occurs. The maximum moves from one part of the membrane to another as the frequency of the sound wave is varied, and it is the detection of the place where the amplitude is largest that constitutes frequency discrimination. Each axon, or group of axons in the auditory nerve, carries information about a particular frequency, and the amplitude of any particular frequency carried by the corresponding axon is coded by the frequency of pulses. (Since the response of a sensory cell is usually

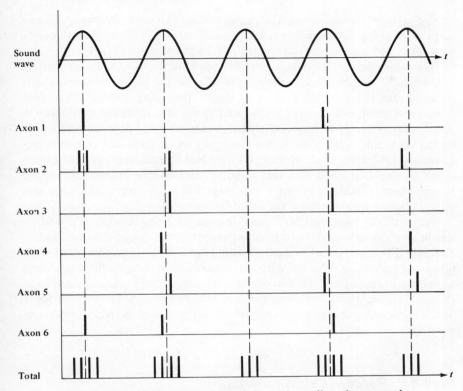

Figure 7.9 Schematic illustration of how the volley theory works. Although any particular axon can only respond once every few cycles, the sum of a number of axons reproduces the incident frequency.

logarithmic with stimulus intensity, it appears that the decibel system of measuring loudness is a very natural one.) This theory, called *place theory*, would have us believe that the auditory system is equivalent to a spectrum analyzer that measures each Fourier component of a complex sound and transmits the amplitude along separate channels to the brain.

Place theory is supported by a long and difficult series of experiments with human cochleas. Von Bekesy drilled tiny holes in them at various points along their lengths and mounted miniscule mirrors on the basilar membrane. He then used a very carefully designed optical system (there were no lasers then) to observe the motion of those mirrors as the Basilar membrane vibrated. He found that over the range of frequencies from about 100 to 10,000 Hz the basilar membrane vibration amplitude was maximum at some particular point and less than this maximum on either side of this point. Furthermore, when the sound wave frequency was shifted, the location of the maximum amplitude of vibration shifted, and he was able to map the regions of the basilar membrane that responded most strongly to each frequency. These maps were fairly reproducible from one cochlea to another. Von Bekesy's work provided a rather solid experimental basis for place theory.

The major drawback to the theory derives from the mechanical properties of the basilar membrane. It is simplest to imagine the basilar membrane as a series of taut strings transverse to its length, providing position-dependent resonances much like a harp. Unfortunately for this idea, the membrane is under no tensions at all but simply floats in the cochlear fluid. Furthermore, its thickness and rigidity vary in such a way as to make the resonant frequencies vary by about a factor of 100 from end to end, whereas our hearing range is nearly 10 times as large. Finally, place theory cannot explain the binaural beats described by Oster. It is clear that place theory, like telephone theory, cannot be a complete theory of hearing either.

Place theory has the interesting feature of being compatible with a sharpening mechanism derived from lateral inhibition (to be discussed later). The actual mechanism of human hearing is probably some combination of place and telephone theories, with some modifications to both. Many ideas have been proposed and worked out in some detail, but so far none has been adequate to explain all the phenomena of human hearing. To this day, some of the simplest questions we can ask about auditory perception are unanswered.

Part 5: Perceptual Mechanisms

In this chapter we have discussed the production, properties, propagation, and detection of sound. We now turn to the much more difficult questions of perception and interpretation. Most of these questions have no answers and/or are beyond the scope of this book. Nevertheless, they can be introduced for discussion, especially those few cases where we do have some understanding of the phenomena.

First, we consider the curious phenomenon of lateral inhibition. It has long been known that there are many ways nerves can form junctions (synaptic junctions) and that some of them are inhibitory. That is, if two axons feed into a single ganglion cell, it is possible that the ganglion's response to a fixed stimulus from one of the axons could be reduced by a stimulus from the other axon. In this case we say that the other axon has an inhibitory effect.

In order to understand how inhibition affects perception, we consider an array of receptors as shown in Figure 7.10(a). Suppose that each cell in the array could inhibit the response of its nearest neighbor, but only by some fixed amount of the original cell's response. If the array is subject to a non-

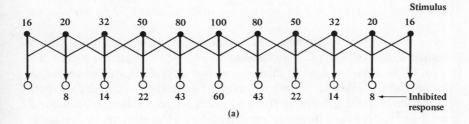

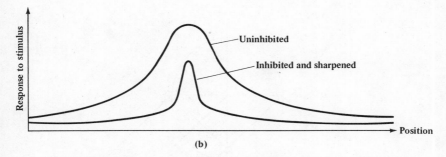

Figure 7.10 Schematic illustration of lateral inhibition. In (a) an array of cells is subjected to stimuli of relative magnitude given by the top row of numbers. Since each cell sums the stimulus incident on its two nearest neighbors and is inhibited by 25 % of that amount, the second row of numbers represents the relative response of the array of cells. (b) is a plot of the stimulus and response to it versus spatial extent along the array of cells.

uniform stimulus that has a maximum located at one of the cells, that cell will exert a stronger inhibitory effect on its neighbors than they will exert on it, resulting in an enhancement of its relative strength. In the numerical example of Figure 7.10, we assume each cell can inhibit the response of its neighbors by 25%. The cell whose uninhibited response would be 80 is reduced by 25 % of 50 ($=12.5$) and 25 % of 100 ($=25$) so that its net response is only 42.5 (similarly for the other cells). We notice that the center cell that had enjoyed only a 25 % higher level of response to distinguish it as being at the maximum before inhibitory effects is now elevated to a 40 % higher level over its neighbors. Inhibition can act in this way to sharpen sensory perceptions.

Suppose the detectors were hair cells along the basilar membrane. In order to explain the precision of our tonal discrimination, the place theory requires that we determine which part of the basilar membrane is vibrating with maximum amplitude to within a small fraction of a millimeter. It is clear that the sharpening effects of the lateral inhibition scheme described here can be of great help, especially if there is more than one stage of the process. The effects of inhibition are much more important in describing visual phenomena and will be considered in more detail when we take up that topic.

The next subject to be considered is the localization of sound. It is clearly a desirable quality of the auditory system to be able to locate the source of a sound. If the sound is a short click, its arrival at one ear will be earlier or later than at the other ear simply because it has further to travel. In Figure 7.11 we see that sound coming from a distant source located at an angle θ away from the forward direction must travel an extra distance $x = d \sin \theta$.

Top view

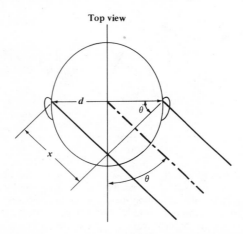

Figure 7.11 Top view of a human head and its two ears showing coordinate system for discussion of directional sensitivity of hearing.

This requires a time interval $\Delta t = x/v$ and, for $v = 320$ m/sec, the maximum value of Δt ($\sin \theta = 1$) is a little less than 1 msec. Experiments have shown that if two sequential clicks are separately presented to the two ears, they are distinguishable from a single click only if they are separated by at least 2 msec. Otherwise, they are interpreted as a single click coming from some particular angle corresponding to the appropriate time delay. Furthermore, since our perception of the direction from which a sound comes is accurate to about 10°, it is clear that we can distinguish intervals as small as 40 μsec (at $\theta \simeq 45°$) and interpret them as directional effects. The 40-μsec interval is to be compared with the 1 msec characteristic time constant of axons.

If the sound whose direction we wish to detect is not a click but is continuous, there is other information that our ears may use. The sound may be louder in one ear than in the other by virtue of the head blocking some of the sound. This will occur only for high-pitched sounds whose wavelengths are small compared to the head size, i.e., those that have frequencies larger than about 5000 Hz. The head-shadowing effect is important for detection of stereo separation and it is clear why a good stereo must have accurate reproduction (without phase shifts) of the high frequencies. For low-pitched sound waves, there may very well be a detectable phase difference between the sounds heard at each ear. Although the phase difference technique of spatial discrimination may lead to ambiguities, it certainly could augment other sources of information about the location of the origin of the sound. Many experiments have been done, especially with blind people, and there is an extensive literature on this subject. Naturally it is of considerable interest to the military.

Another example of hearing experiences that is also quite complex is referred to as "the cocktail party effect". In a noisy environment such as one might find at a cocktail party, it is possible to listen to a conversation between two people and ignore the interference from several other conversations going on at the same time. It is then possible, without any change in the loudness of the conversations for the listener to "tune out" one conversation and begin to listen to another. This seems to require only a conscious effort with no physical changes. That means that all the information from several conversations is always going to our brain, but somehow we are able to sort out one particular conversation at the expense of the others. The fact that we can readily switch from one to another is nothing short of amazing.

Still another example of the complex hearing system is our ability to perceive certain combinations of tones as chords, and certain sequences as tunes. We do not know or understand the process used by the brain to sort out the various sequences which allow identification of a Beatles tune or a Beethoven theme after only a few notes. A series of interesting experiments in musical illusions has been described by Deutsch.

There are several more subtle aspects of hearing that have not yet been discussed. These involve questions that require not only perception of sound but also interpretation of what is being heard. For example, we are familiar with the musician who can enter a room with music playing and, within a few seconds, identify the composer. He might say "that's Bach, but I have never heard the piece before." In another example, a musician listening to a recording might be able to identify the conductor of the orchestra. In each of these cases, there is something more involved than the simple perception of sound and tonal discriminations. The complicated aspects of style, technique, art, learning, and other functions of the human mind come into play in ways that are not well understood.

APPENDIX H
THE MUSICAL SCALE

In Western music the scale has 12 notes, 7 of which are given the names of the first seven letters, A to G. The other five lie between these and can be called either sharps or flats, depending on the musical circumstances. Figure H.1 shows two octaves of the piano keyboard.

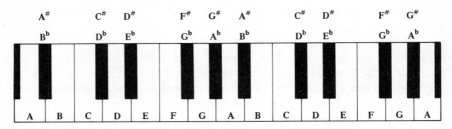

Figure H.1 The piano keyboard.

In order to play the major scale, do re mi fa sol la ti do, as we are accustomed to hearing it, one can start on C and play eight consecutive white notes. If the major scale starts on any other key, it must contain black notes. The key of G contains F# (read *F-sharp*), the key of D contains both F# and C#, etc. One sharp is added each time the key is advanced by one musical fifth (fifth note in the scale). The note sharped is always the seventh note in the scale, so it too progresses in fifths. A similar sequence holds for the use of flats.

The frequency interval between the notes depends on certain choices and compromises. The text describes the even-tempered scale that has equal ratios of frequencies of adjacent notes. Western music offers a variety of other musical scales as well. The Pythagorean scale has these ratios equal to the ratio of small whole numbers as follows: C = f, D = $(\frac{9}{8})$f, E = $(\frac{5}{4})$f, F = $(\frac{4}{3})$f, G = $(\frac{3}{2})$f, A = $(\frac{27}{16})$f, B = $(\frac{15}{8})$f, and C′ = 2f. This scale preserves the harmonious ratios of small whole numbers, but the frequency intervals between notes are not all the same. For example, $\frac{5}{4}$ is not the same as $(\frac{9}{8})^2$. The Pythoragorean scale does not lend itself easily to transposition between keys (scales beginning on different notes) for this reason.

NonWestern music has very different scales. There are as many as 20 notes to the octave (factor of 2 in frequency), with various spacings in some scales, and there are geographical and cultural variations of these. Some Chinese music is based on a scale of many notes but uses only five in any one piece of music. Harmony and dissonance, as well as other asthetic characteristics of music, are both cultural and subjective distinctions.

Questions and Problems for Chapter 7

1. Show that Eq. 7.5 is the correct solution to Eq. 7.3.

2. Consider the case where ω_0 is much larger than γ in Eq. 7.6 and expand the radical to show that the frequency change is indeed small. How many oscillations occur during the time it takes the amplitude to decay to about one-tenth of its original value?

3. Derive Eq. 7.8 by direct substitution of $x = A \exp[i(\omega t + \phi)]$ into Eq. 7.7.

4. Solve Eq. 7.8 and verify that Eqs. 7.9 are correct.

5. Use Eq. 7.9 to show that the width of the curve at the half-maximum point is given by $\Delta\omega = \gamma(1 + \gamma/\omega)$ for the case where γ/ω is small compared to unity.

6. Very often the energy of a vibrating system depends on the square of the amplitude. Calculate the maximum amplitude squared (energy) of a driven oscillator and the half-height points on the energy (amplitude squared) curve. What is the width of the energy curve?

7. Combine Eqs. 7.11 as described in the text and derive the differential equations for the Y's.

8. Find the frequency of sound waves whose wavelengths are comparable to the size of a room, a piece of furniture, and your head.

9. The threshold of auditory damage is $160 \, \mathrm{dB} \cong 2 \times 10^{-4} \, \mathrm{dyne/cm^2}$. Atmospheric pressure is 50 times greater than this. Explain why the ear is not normally damaged by the atmosphere.

10. Give an intuitive reason why the energy of a sound wave depends on the square of the pressure.

11. Since pressure fluctuations of sound raise and lower the pressure by only a small fraction (1 % at most), it seems that all sounds should carry about the same energy. Why is this not correct?

12. The frequency of "Concert A" is 440.0 Hz. Use a calculator to determine the frequency of middle C ($\cong 256$ Hz) and then the E and G above it. How close are these tones to perfect Pythagorean harmonies?

13. How can we possibly be sensitive to vibrations whose amplitude is smaller than the diameter of one atom?

14. Show that $p_0 = 2 \times 10^{-4} \, \mathrm{dyne/cm^2}$ corresponds to an energy of $10^{-6} \, \mathrm{watt/cm^2}$ (good exercise in keeping units straight).

15. Discuss some of the most important theories of hearing. Describe their most important advantages and disadvantages in the light of known properties of the auditory system.

16. Pronunciation of each of the vowel sounds of the English language at the same pitch (fundamental frequency) will all sound different. An "ooo" sound is quite distinguishable from an "eee" sound even if both are sung at middle C. Explain

179

why. Use diagrams to explain your answers. Show how the waveforms of the sounds might look.

17. Exactly what is a decibel? What is it used for? If a 20-dB sound is made twice as loud, how many decibels will it be?

18. One of the functions of the middle ear is to increase the pressure provided to the inner ear when a sound is heard. Why doesn't this violate conservation of energy?

19. Why must a good stereo system have accurate reproduction for both highs and lows? What aspects of musical perception are carried by which end of the audible frequency spectrum?

Bibliography

1. A. H. BENADE, *Horns, Strings and Harmony* (Doubleday, Garden City, N.Y., 1960). (A wonderful discussion of various topics in the physics of music.)

2. A. H. BENADE, *Fundamentals of Musical Acoustics* (Oxford, New York, 1976). (A solid textbook on sound, music, and musical instruments.)

3. A. G. BOSE, "Sound Recording and Reproduction". (An excellent two part article in the June and July/August, 1973 issues of Technology Review) (MIT Alumni Magazine).

4. D. DEUTSCH, "Musical Illusions" Scientific American, October, 1975.

5. F. GELDARD, *The Human Senses* (John Wiley and Sons, Inc., New York, 1972). (A thorough, careful treatment of a great many sensual phenomena and theories.)

6. J. JEANS, *Science and Music* (Dover Publications New York, 1968). (A careful discussion of physics and music.)

7. J. JOSEPHS, *Physics of Musical Sound* (D. Van Nostrand, Princeton, N.J., 1967).

8. G. OSTER, "Auditory Beats in the Brain," *Scientific American*, 94 (October 1973).

9. J. ROEDERER, *Physics and Psychophysics of Music* (Springer-Verlag, New York, 1974). (A good discussion of musical acoustics and various auditory phenomena.)

10. T. RUCH and J. FULTON, *Medical Physiology and Biophysics* (W. B. Saunders Co., Philadelphia, Pa., 1960). (A standard physiology text.)

11. E. SELKURT, ed., *Physiology* (Little, Brown and Company, Boston, Mass., 1971). (A standard physiology text.)

12. W. VAN BERGEIJK, J. PIERCE, and E. DAVID, *Waves and the Ear* (Doubleday, Garden City, N.Y., 1958). (A wonderful discussion of various topics in the physics of music.)

13. G. VON BEKESY, *Experiments in Hearing* (McGraw-Hill Book Co., New York, 1960). (A description of his Nobel Prize winning work.)

8

LIGHT, COLOR, AND VISION

Part 1: Electromagnetic Theory and Light

In the early part of the nineteenth century, natural science included many separate and distinct topics; among them were electricity, magnetism, and optics. By 1850, the work of Faraday, Henry, and others had merged the sciences of electricity and magnetism, and the work of Young and Fresnel had laid the foundation of wave optics. In 1867, James Clerk Maxwell made one of the most astonishing discoveries of modern science. In the course of providing a consistent mathematical description of the relation between electric and magnetic phenomena, he discovered that it was possible to propagate an electromagnetic disturbance through space and that the speed of propagation was nearly exactly that measured by Fizeau and others to be the speed of light! He was immediately convinced that light is an electromagnetic phenomenon and that he had found a connection between the science of optics and the science of electromagnetism. Eight years after Maxwell's death and twenty years after his discovery, electromagnetic waves were observed by Heinrich Hertz. Today we treat light as an electromagnetic disturbance having all the properties of classical waves.

We begin our review of the properties of light waves by writing Maxwell's equations:

$$\text{div } \vec{E} = 4\pi\rho, \tag{8.1a}$$

$$\text{div } \vec{B} = 0, \tag{8.1b}$$

181

$$\text{curl } \vec{E} = \left(\frac{-1}{c}\right)\left(\frac{\partial \vec{B}}{\partial t}\right), \tag{8.1c}$$

$$\text{curl } \vec{B} = \left(\frac{1}{c}\right)\left(\frac{\partial \vec{E}}{\partial t}\right) + \left(\frac{4\pi}{c}\right)\vec{J}. \tag{8.1d}$$

Equation 8.1a is Gauss's law and depends on the $1/r^2$ nature of the Coulomb field. Equation 8.1b is the equivalent statement for magnetic fields, except that since magnetic monopoles have not been observed, the right-hand side is zero. Equation 8.1c embodies Faraday's and Henry's law of electromagnetic induction: note the negative sign (Lenz' law) that requires that the induced field be such as to oppose the current change. This equation would have to be modified by the addition of another term if the existence of the Dirac magnetic monopole were confirmed. Equation 8.1d contains Ampere's law (second term on the right-hand side) as well as the famous Maxwell displacement current (first term on the right).

These equations were written by Maxwell in order to obtain consistency and symmetry in the mathematical description of electromagnetism. His version, of course, did not have the speed of light, c, explicitly as these do. Rather they contained the permeability ϵ_0 and permitivity μ_0 of free space. We have used the relation $c^2 = 1/\epsilon_0\mu_0$ in Equations 8.1. By taking the curl of the second pair of Maxwell's equations and using well-known vector identities, one can derive the (one-dimensional) electromagnetic wave equations:

$$\frac{\partial^2 \vec{E}}{\partial t^2} = c^2 \frac{\partial^2 \vec{E}}{\partial x^2} \tag{8.2a}$$

and

$$\frac{\partial^2 \vec{B}}{\partial t^2} = c^2 \frac{\partial^2 \vec{B}}{\partial x^2}. \tag{8.2b}$$

Equations 8.2a and 8.2b predict the existence of electromagnetic waves propagating in free space at the speed of light but imply no limit to their wavelength or frequency. Although our eyes are sensitive to electromagnetic (EM) waves over a very narrow range of wavelengths and frequencies called visible light, we have built instruments that allow observation over 20 orders of magnitude in frequency (see Figure 8.1). Within the narrow visible range of a factor of 2 (one octave) of EM waves we can perceive the pleasures of art and scenery along with learning most of what we know. This narrow frequency range is to be contrasted with the range of a factor of nearly 1000 for audible sound. It is indeed difficult to imagine what the world would look like if we could see EM waves whose frequency range varied over a factor of 1000.

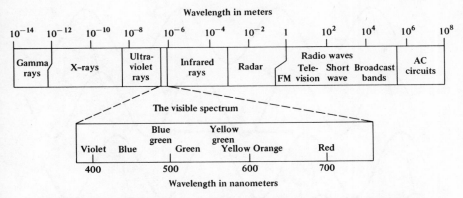

Figure 8.1 The known electromagnetic spectrum spans more than 20 decades of frequency and wavelength. Maxwell's equations put no limit on the extent of the spectrum.

Part 2: Review of Geometrical Optics

Maxwell's wave equations have solutions

$$\vec{E} = \vec{E}_0 \cos 2\pi\left(ft - \frac{x}{\lambda}\right) \tag{8.3}$$

which represent plane waves traveling in the x direction; that is, at every point in the yz plane of a particular value of x, the electric field has the same value. As one moves in the x direction, the field varies sinusoidally with wavelength λ, and if one sits at rest at some point, the field varies sinusoidally in time at frequency f. The speed of propagation v is given by

$$v = f\lambda \tag{8.4}$$

because it is clear that if we moved at that speed $x = vt = f\lambda t$, the field would be constant and we would be "riding" the wave at its own speed.

It is often convenient to make a pictorial representation of a plane wave by simply indicating the direction of propagation with a single line or arrow (see Figure 8.2). This line is always drawn perpendicular to the wave crests and is called a *ray*. Although we perform many calculations using this simple pictorial device, we must always remember that it is limited in utility and that it is not reality. The ray optics pictures we shall use are very convenient and helpful, but we must always be careful to avoid pitfalls in drawing conclusions.

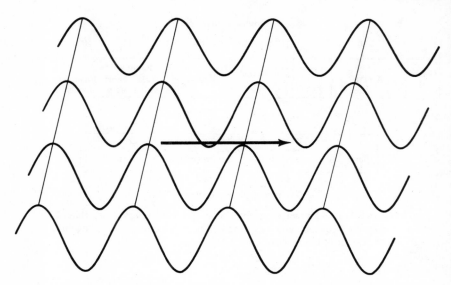

Figure 8.2 A coherent wave consisting of crests and troughs is represented by a single ray that points in the direction of propagation.

When light travels in a medium other than free space, its speed is different from c. The speed in any material is determined by its index of refraction n and is given by

$$v = \frac{c}{n}. \tag{8.5}$$

We now consider what happens to plane light waves when they change speed on traversing the boundary between two media. If the angle of incidence is 90° as shown in Figure 8.3(a), the slowing of the waves is manifest as a reduction of wavelength. It is obvious that the frequency cannot change since the boundary can neither create nor destroy waves. In order to satisfy both Eqs. 8.4 and 8.5, it is clear that the wavelength must be

$$\lambda = \frac{\lambda_0}{n} \tag{8.6}$$

where λ_0 is the wavelength in vacuum.

If the angle of incidence of the plane waves on the boundary is not 90°, then the waves must bend as shown in Figure 8.3(b) if Eq. 8.6 is to be satisfied. There is no other way to satisfy all the requirements above.

In Figure 8.4 we see the crests of two successive waves impinging on a boundary xx' between two media labeled 1 and 2. We define the angle θ_1 between ba' and the perpendicular to xx' as the angle of incidence and see immediately that it equals angle baa'. Similarly, the angle of refraction θ_2 between ac and the perpendicular to xx' is equal to angle $aa'c$ as indicated.

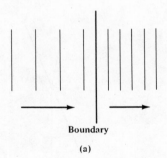

Boundary

(a)

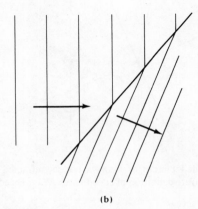

(b)

Figure 8.3 Plane waves are refracted at a boundary. Here the waves propagate into a medium of slower speed at a right angle to the boundary (a) and at an oblique angle (b).

The two right triangles in Figure 8.4 share the common hypotenuse aa', so we can write

$$\frac{(ba')}{\sin \theta_1} = aa' = \frac{(ac)}{\sin \theta_2}. \tag{8.7}$$

We note that ba' and ac are the wavelengths in each of the media and use Eq. 8.6 to find

$$n_1 \sin \theta_1 = n_2 \sin \theta_2 \tag{8.8}$$

which is the famous Snell's law for refraction. We commonly use this equation to describe the bending of light rays as they pass boundaries between different media.

It is rather simple to show that a spherical boundary between two media will cause light rays to be focused to a single point called the focal point as shown in Figure 8.5(a) and that a simple lens [Figure 8.5(b)] will do the same.

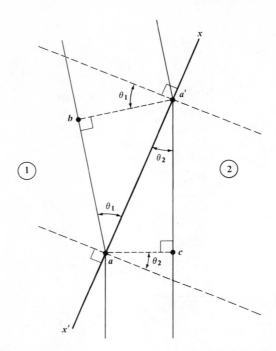

Figure 8.4 Detailed picture of two wavecrests refracted obliquely at a boundary to illustrate definition of coordinates. Wave speed in medium 2 is slower.

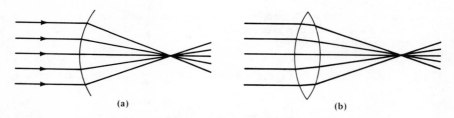

Figure 8.5 Refraction at a curved boundary illustrated with the ray picture. In (a) there is a single boundary and in (b) the two boundaries form a lens.

The distance from a single curved surface to the focal point for plane waves is called the *focal length f* and is given by

$$f = \frac{n_2 R}{n_2 - n_1} \tag{8.9}$$

where n_2 is the index of refraction on the convex side of the boundary whose radius of curvature is R (R is positive). For a simple lens made of material having index of refraction n_2 and immersed in a medium of index n_1, the focal length f is given by

$$\frac{1}{f} = \left(\frac{n_2 - n_1}{n_1}\right)\left(\frac{1}{R_a} - \frac{1}{R_b}\right) \qquad (8.10)$$

where R_a and R_b are the radii of curvature of the two lens surfaces measured from the same side of the lens. A double convex lens has a positive focal length. The combination of Eq. 8.9 and 8.10 allow us to relate the distance between a thin lens and an object s to the distance between the lens and the image s'. After considerable calculation, we find

$$\frac{1}{s} + \frac{1}{s'} = \frac{1}{f}. \qquad (8.11)$$

Part 3: Physiological Optics

We are now prepared to investigate the image forming properties of the optical system in the human eye. Since the front-to-back diameter of the eyeball is about 22 mm, we see that the desired image distance s' in Eq. 8.11 is 22 mm for all values of s, the object distance. Furthermore, the range of values of s over which we can focus clearly is from infinity to about 20 cm. The eyeball must be able to accommodate its focal length from 22 mm (infinitely distant object) to about 19.5 mm (20-cm distant object).

The lens of the eye is a partially flexible structure that can be distended by its supporting muscles. The change in shape changes its focal length and permits accommodation for the focusing of both near and far objects (but not simultaneously). The relaxed lens is held in its flattest shape by a set of ligaments and muscles around its periphery. In order to focus closer objects, the lens is thickened by the action of these muscles. Anatomical measurements show that the relaxed radii of curvature of the lens (index of refraction = 1.413) are 6 and 10 mm and that it is immersed in a fluid of index of refraction 1.336. According to Eq. 8.10 its (minimum) focal length is therefore 65 mm.

We next consider the combined focusing effect of the lens and cornea. The radius of curvature of the outer surface of the cornea is about 7.7 mm and its index of refraction is about 1.376. We use Eq. 8.9 to find its focal length to be about 28 mm because it's in contact with air. When two lenses of focal length f_1 and f_2 are placed close together, their imaging properties can be described by a single focal length F given by

$$\frac{1}{F} = \frac{1}{f_1} + \frac{1}{f_2}. \qquad (8.12)$$

Equation 8.12 can be used to find the focal length of the relaxed eye, $F = 19.6$ mm. The distance from the midpoint between the lens and cornea to the retina is about 20 mm. Therefore, the relaxed eye focuses objects at large distances onto the retina. The shortest focal length that the lens can achieve is about 45 mm, which means that the closest object that can be focused onto the retina is about 15 cm from the eye. Most people's eyes cannot achieve this minimum, however, and the usual closest distance is 18 to 20 cm.

Note that most of the refraction responsible for this focusing is done by the cornea and not by the lens. This is because the cornea is in contact with the air and there is a much larger change in the index of refraction across its boundary than across those of the lens. We have great difficulty forming clear images under water because the cornea is then in contact with the water whose index of refraction (1.33) is much closer to that of the cornea. Wearing a diver's mask alleviates this condition by providing an air-cornea interface to allow for focusing.

If a lens of focal length f' is placed in front of the eye, the resulting optical system will have a focal length F' given by

$$\frac{1}{F'} = \frac{1}{F} + \frac{1}{f'} \tag{8.13}$$

where F is the focal length of the unaided eye. This equation is, of course, the same as Eq. 8.12. In many cases, F is not the right size to form clear images either for close objects or for distant ones, and it is necessary for an individual to wear eyeglasses. It should be clear how to calculate the focal length f' of the corrective lenses needed for various kinds of visual difficulties.

It is usually the job of an optometrist (not an opthalmologist, who is a physician, nor an optician, who makes the glasses) to determine what kinds of corrective lenses are needed to remedy a particular focusing problem. Many people (especially older people) have limited accommodation in both stretching and flexing of the lens and may require help in forming images of both near and far objects (see Figure 8.6). *Bifocals* are lenses which are made

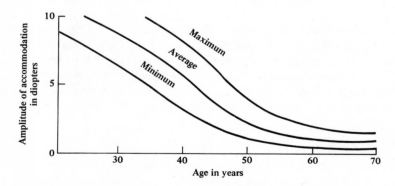

Figure 8.6 Plot of visual accommodation range versus age (humans).

from two pieces of glass which are parts of two lenses of different focal lengths. By looking through the upper lens (usually of long focal length, often negative or diverging) it is possible to focus the image of distant objects and by looking through the lower (usually shorter focal length) lens it is possible to focus the image of a close object. The "strength" of an eyeglass lens is measured in diopters $= 1/f'$ where f' is the focal length measured in meters. Most eyeglass prescriptions are in the range of one to four diopters, which means the lenses have focal lengths in the range from 1 m to 25 cm.

The two most common focusing defects that are corrected by glasses are myopia (nearsightedness) and hyperopia (farsightedness). Each of these cases is illustrated in Figure 8.7 along with a diagram showing the effect of a corrective lens.

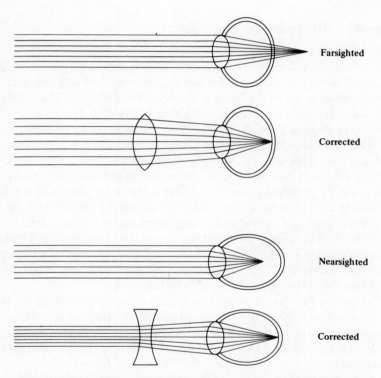

Figure 8.7 Ray diagrams showing how eyeglasses correct improper focusing for both nearsighted and farsighted eyes.

When a lens forms an image at distance s' of an object at distance s and size h, the image has a size $h' = -hs'/s$ where the negative sign means the image is inverted. We define the magnification of the system m as

$$m = \frac{h'}{h} = -\frac{s'}{s}.$$ (8.14)

When we want to have a good look at something small, we bring it as close to our eye as possible and still form a good image. By reducing s in Eq. 8.14 to its minimum (about 18 cm) we achieve the largest value of m (about 0.12 since $s' = 22$ mm). By using an auxilliary lens of focal length f_1, however, it is possible to reduce the focal length of the lens-eye system as with eye-glasses to a substantially lower value and therefore reduce s below 18 cm. This is the basic principle behind a magnifying glass as well as the eyepieces of telescopes and microscopes. The magnification from such a lens is approximately (see Section 9-4)

$$m = \frac{18 \text{ cm}}{f_1}.$$ (8.15)

Part 4: Anatomy of the Eye

Our light-sensitive organ, the eye, is a very complicated device. It has a variety of control mechanisms for aiming, focusing, and aperture control. It conveys information to the brain through a thick bundle of fibers called the *optic nerve*. Its basic structure is much like a camera in the sense that it has a refractive element in the front that is separated from a photosensitive surface where an image is formed. The similarity ends there, however, for as we shall see, the retina is in no way like photographic film.

Figure 8.8 is a schematic vertical cross-sectional view of the eye which shows that the principle refractive surface is the outer surface of the cornea and that it has greater curvature than the rest of the eyeball. All the other boundaries along the optical path are between media whose index of refraction differ by only a few percent. At the cornea, however, there is a boundary between media of index 1.38 and 1.0 (the air). The presence of a large difference of refractive index at the boundary supports our previous discussion that image formation is done primarily by the cornea.

The pupil of the eye is defined by the *iris*, which lies between the cornea and lens and also gives the eye its characteristic color. The pupil can be enlarged or constricted through the action of a *sphincter muscle* that responds to the total average light energy entering the eye. The iris is immersed in the *aqueous humor*, a watery fluid that fills the volume between the cornea and lens.

The eyeball itself is filled with a clear jellylike mass called *vitreous humor*. This fluid supports the round shape of the eye. It bathes the back side of the lens that is held in place on the inner wall of the eyeball by the *ciliary muscles*. These are the muscles that stretch and relax to allow the lens to change its shape and therefore its focal length.

The boundary of the eyeball has several layers (*sclera, choroid,* and *retina*) but we shall be concerned only with the innermost layer, the retina. It forms

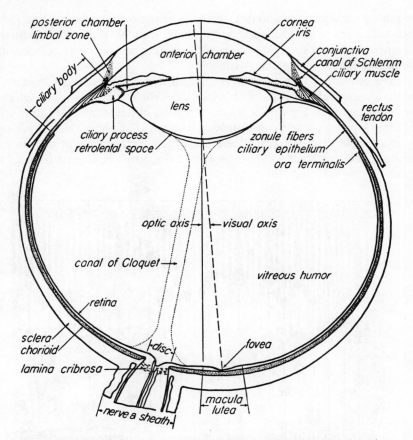

posterior chamber
limbal zone
ciliary body
anterior chamber
cornea
iris
conjunctiva
canal of Schlemm
ciliary muscle
lens
rectus tendon
ciliary process
retrolental space
zonule fibers
ciliary epithelium
ora terminalis
optic axis
visual axis
canal of Cloquet
vitreous humor
retina
sclera
chorioid
lamina cribrosa
disc
fovea
nerve a sheath
macula lutea

Figure 8.8 Cross section of the eye viewed from the top. From Ruch and Fulton, *Medical Physiology and Biophysics*, 18th ed. W. B. Saunders Co., Philadelphia, 1960.

a continuous surface on the inner boundary that is broken at only two points. One of these is the point where the optic nerve leaves the eyeball. Since there is no place for any receptor cells there, we have a small blind spot just to the nasal (inner) side of the center of our field of view. The other break in the uniformity of the retina is a small pit, less than 1 mm in diameter, called the *fovea*. This extremely important part of the visual system will be discussed in detail.

The fovea is the site of our best visual acuity and all color vision. The structure is readily distinguishable from the rest of the retina for a variety of reasons. To begin, there are two basically different kinds of cells found in the retina: the *rod cells* are distributed throughout the retina, whereas the *cone* cells are found exclusively in the fovea. Furthermore, although there are 100 million cells in the retina and only 1 million axons in the optic nerve, requiring a reduction of 100 to 1, there is a nearly one-to-one correspondence between foveal cells and optic nerve fibers. Finally, most of the retina is characterized by a complicated array of interconnecting nerve cells but this complexity is reduced at the fovea.

The retina is a very complicated organ. It is composed of at least six or seven anatomically distinguishable layers of cells whose individual functions are quite complex. Figure 8.9 is a cross section of the retina that shows some of these layers and their constituent cells. We shall see later that the multiple interconnections of the retinal cells perform a certain amount of data processing, and this leads to the notion that the retina is an extension of the brain.

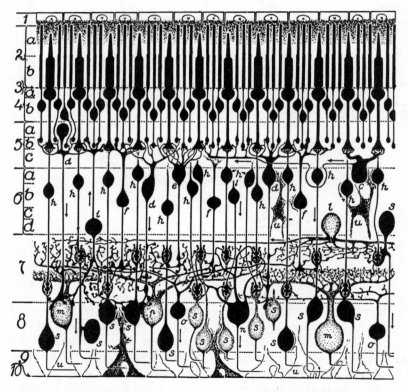

Figure 8.9 A schematic diagram of the cross section of the human retina is shown magnified several thousand times. Light is incident from the bottom and passes through several distinctly visible layers before it is detected in the outer segments in layer 2. Layer 4 contains the inner segments of the rod and cone cells. Layers 5 through 8 contain many synaptic junctions where significant data processing occurs, and layer 9 contains the origins of the optic nerve fibers. From Ruch and Fulton, *Medical Physiology and Biophysics*, 18th ed, W. B. Saunders Co., Philadelphia, 1960.

The optic nerve leaves the back of each of the eyeballs and splits into two branches. Those axons from receptors in the left (outer or temporal) side of the left eye join with those from the left (inner or nasal) side of the right eye and proceed to the left lobe of the visual cortex at the back of the brain. The right lobe is similarly fed by axons from receptors in the right side of each eye. The crossing of the parts of the optic nerve from the nasal part of each eye is called the *optical chiasma* and occurs just behind the eyes. The separation of the right and left fields of view rather than the images from the right and left eyes raises questions about the basis of stereoptic vision and certain phenomena associated with depth perception. These must be considered very carefully to avoid falling into logical inconsistencies.

Part 5 : Visual Acuity

Visual acuity is the ability to see fine detail and discriminate small objects. One must take care in any quantitative descriptions or measurements of acuity because there are so many ways in which it can be measured. Experimentors may use tests that require detection, identification, resolution, or localization of an image as well as any combination of these. Furthermore, the detection criteria may be at a 50% accuracy threshold or at some other arbitrary level. It is necessary to be extremely careful in designing an acuity test: we can detect light from an arbitrarily small source if it is bright enough.

There are a number of standard tests to measure acuity. The most common of these, the Snellen chart, is often seen in many doctors' offices. The subject views an array of letters of various sizes and is asked to read them, one eye at a time. Another fairly common acuity test, called the Landholt C test, requires the subject to determine the direction of the opening of a carefully constructed letter C (see Figure 8.10). There are a variety of other

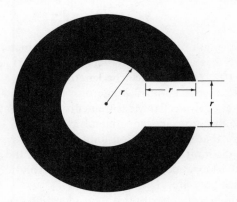

Figure 8.10 Drawing of the Landholt C.

tests involving gratings (sine or square wave) and other geometrical shapes. The results of acuity measurements in each of these experimental techniques vary somewhat from one to another, but in general the visual acuity of the normal average eye is found to be a little less than about 1 minute of arc or about 2×10^{-4} radian. This corresponds to the diameter of a human hair held at arm's length. The vernier acuity test gives a result that is different from this average, and we shall discuss it separately later.

There is a reasonably good physical as well as physiological basis for this limit to visual acuity. The average retinal cell has a diameter of about 1 micron (10^{-3} mm) and is located about 22 mm from the lens-cornea optical system. It therefore subtends an angle of about 5×10^{-5}. Since it takes more than one cell to determine the presence or absence of a boundary in an image as required by an acuity test, we might expect that the minimum size visual field that could be used to detect a boundary would be 2 or 3 microns corresponding to an angle of about 10^{-4}. The finite size of the cells therefore imposes a retinal mosaic limit to visual acuity that corresponds very closely to the observed capability of the human eye. This simple argument cannot be pushed too far, however, because it assumes that the brain is somehow aware of the location of every single cell in the fovea. (The fovea is the only part of the retina that has acuity as fine as 10^{-4}.) Clearly this information cannot be available to the visual cortex because the foveal cells are subject to all the irregularities of biological growth in terms of size, shape, location, and viability.

Let us look at another limit to visual acuity. If the eye were an ideal optical system with perfectly shaped spherical surfaces and made from perfectly clear materials, the wave nature of light would impose a limit on the precision of focusing light. This *diffraction limit* is a fundamental characteristic of all optical systems and is usually described in terms of the angular width of the spot formed by the optical system when it focuses parallel light. The angular width θ is

$$\theta = \frac{1.22\lambda}{D} \tag{8.16}$$

where λ is the wavelength of the light and D is the diameter of the aperture. For the human eye, λ is about 5×10^{-5} cm (green light) and D is typically 4 mm giving $\theta \cong 10^{-4}$ radian. Since the length of the eye is about 22 mm, this corresponds to a spot about 2.2 microns in diameter—just about the same as the diameter of two cells! Measurements confirm that parallel light is focused into a disc that does indeed subtend an angle of 10^{-4} radian. It is astounding to learn that the diffraction limit to human visual acuity is just about the same as the limit imposed by the finite size of the retinal cells. This fact should lead us to extended speculation about evolution and about why our eyes are the size they are.

There are a number of other optical considerations we might discuss in connection with visual acuity. The problems of chromatic aberration from the optical system are largely alleviated by the presence of a yellowish pigment called *macula lutea* in the fovea. This colorant partially filters out the light of longer and shorter wavelengths that contribute most strongly to chromatic aberrations. Spherical aberration is surely present in the visual system, but its effects are also minimal. This is because the pupil diameter of typically 4 mm is considerably smaller than the 22-mm focal length of the eye. We use only the central portion of the lens and cornea, which are the areas least subject to the effects of spherical aberrations.

It is possible that inhomogeneities and impurities in the ocular media would scatter light and thereby deteriorate the optical image and reduce visual acuity. The effect of scattering is considerably alleviated by the directional sensitivity of the receptor cells. Because of their shape, size, and orientation, a number of geometrical and waveguide effects combine to make the cells sensitive only to light that comes from the forward direction. This directional sensitivity, called the Stiles-Crawford effect, is characterized by an angular width of about 10°, which means that the sensitivity of the receptors to light incident at angles larger than this is substantially reduced. One geometrical model of the Stiles-Crawford effect is shown in Figure 8.11. Because of the

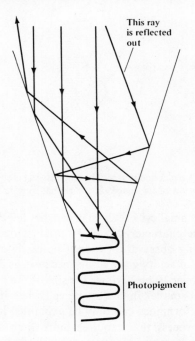

This ray is reflected out

Photopigment

Figure 8.11 Ray optics of a cone. This picture is not accurate for a cone cell whose size is comparable to a wavelength of light. It is necessary to use waveguide modes to describe the light paths.

taper of the cone cell wall, light that is incident at large angles is reflected back out of the cell but light incident at smaller angles is transmitted through the cell to the photosensitive pigment at the back of the cell. Microscope studies of cone cells are consistent with this model. The Stiles-Crawford effect can be considered as a pupillary apodization in certain problems. We have seen that the optical aberrations do not impose substantial limits to visual acuity because the eye is somewhat protected against them. The acuity limits are imposed by the fundamental effects of cell size and diffraction.

It turns out that the nice ideas about visual acuity discussed above do not provide an adequate description of how we see detail for a number of reasons. One of these reasons derives from a particular test of acuity called the *vernier test*. A subject views a line that has been broken and displaced perpendicular to its length by a small amount as shown in Figure 8.12(a). When we try to

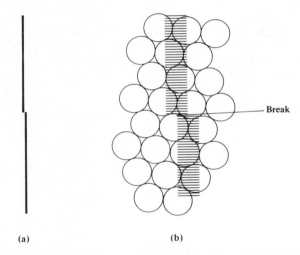

(a)   (b)

Figure 8.12 The vernier acuity test. The test drawing is shown in (a) and a schematic illustration of its image on the retinal mosaic is shown in (b).

determine what is the smallest break that can be detected, we find it to be much smaller than the characteristic 10^{-4} radian. Often it is less than 10^{-5} radian. This astounding observation cannot be easily explained in terms of the simple description of acuity given above because, as Figure 8.12(b) shows, the image is smeared over 10 or 20 times the size of the detected break, which is itself 10 or 20 times smaller than a single cell when imaged on the retina. The vernier acuity performance can only be explained in terms of some sort of signal averaging process that requires more than simple photographic transmission of the image from the retina. In a later section of this chapter we shall have more to say about how visual information is transmitted to the brain.

One of the more modern ways of describing the acuity or fidelity of any linear system is in terms of its transfer function. We provide sine wave stimuli (gratings for an optical system) of various frequencies and ask how the system in question transfers or images these sine waves. We talk of spatial frequencies in the image and measure in terms of cycles per millimeter or cycles per radian. Since diffraction will produce a smearing of any image, it is clear that any optical system has a finite upper limit of spatial frequency f_{max} that can be found from Eq. 8.16:

$$f_{max} = \frac{D}{1.22\lambda F} \qquad (8.17)$$

where F is the focal length. Notice that f_{max} increases with D/F, which is the reciprocal of the f-number of the system. It is clear that optical systems perform best when the f-number is small and for this reason the quality of telescopes and microscopes is generally characterized by their f-number (or the related quantity, *numerical aperture* = N.A. = $nD/2\sqrt{D^2 + F^2}$ = $n \sin \theta_{1/2}$ where n is the index of the medium outside the lens). Oil immersion microscope objectives work by increasing the value of n and thereby increasing the N.A. (See Section 9-4.)

The transfer function (or modulation transfer function, MTF) is given at any frequency by the ratio of the contrasts of sine wave patterns in the image and in the object. For a grating (oscillating intensity) the contrast R is

$$R = \frac{I_{max} - I_{min}}{I_{max} + I_{min}} \qquad (8.18)$$

and the transfer function is simply R_{image}/R_{object}. Since any linear optical system can only deteriorate the contrast, the MTF is always less than unity except at zero spatial frequency (solid light or dark) where it is exactly unity. Since it goes to zero at f_{max}, it clearly has the form shown in Figure 8.13. How

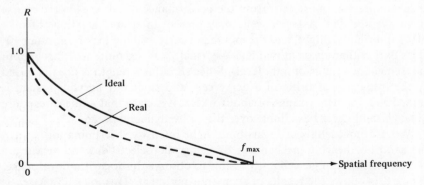

Figure 8.13 The modulation transfer function is a plot of image contrast divided by object contrast (R) versus spatial frequency. It is chosen to be unity at frequency zero and to be zero at the cutoff frequency.

it drops from one to zero is characteristic of the optical system: the MTF for an ideal, diffraction limited imaging system is shown in Figure 8.13.

Various optical aberrations are studied in terms of their effects on the MTF and results of these studies can certainly be applied to the optical system of the eye. One can therefore calculate the MTF for the human eye and compare it with measurements (readily made with live subjects using an altered opthalmoscope). One must be very cautious about drawing conclusions from these studies, however, because the MTF describes the image formed by an optical system and we have already seen in at least one example (vernier acuity test) that images are not transmitted point for point from the retina to the brain. It turns out that studies of the eye's MTF are not of much help in describing the acuity of the human visual system.

Part 6: Intensity, Flicker, and Wavelength Limits of the Visual System

We have discussed the limits of the ability to see detail (visual acuity). We shall now discuss several other limits of the visual system. Detection of light by the receptor cells is basically a quantum process. A single quantum of light causes a transition in the molecules of the visual pigments that results in the initiation of a nervous response and the eventual "seeing" of the light. We might therefore expect that, under ideal conditions, the eye should be sensitive to a single photon, and indeed that is very nearly so. It turns out to be very difficult to make such measurements so that the best results to date have put the threshold at two or three quanta rather than at one in a several msec. period.

The response of the visual system to light of intensity greater than threshold is logarithmic with increasing intensity. It is clear that some form of scale compression must exist from the performance of our eyes in the course of an average day. A given scene may present intensity variations of over 10^6: 1 in bright sunlight, and the minima in early evening may be another 10^4 times darker than those at midday. We usually can see quite well throughout the tremendous range of light levels. Since stimulus strength is usually coded by the frequency of firing of a nerve cell, and since our nervous system is limited to a frequency range of about 1 kHz, it is clear that a linear response system would never be suitable over the entire dynamic range.

We do indeed have a logarithmic light intensity discrimination system just as in our hearing mechanism. By using microelectrodes and measuring the response of a single cell it is possible to observe this logarithmic response. Figure 8.14 shows the results of such an experiment. This logarithmic sensitivity to light intensity has been known from macroscopic measurements of the just noticeable difference (JND) between light stimulus at two different levels. The results of these measurements are embodied in the Weber-Fechner

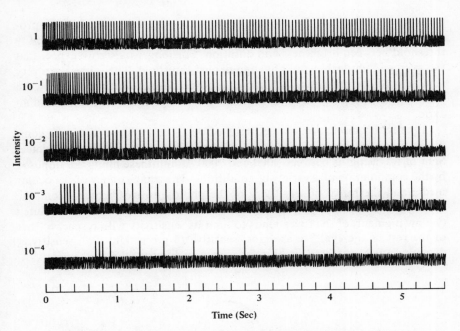

Figure 8.14 Nerve impulse rate triggered by light increases with increasing light intensity but not linearly. Each line shows the response for an increase of a factor of 10 in light intensity. Careful measurements show that the response is indeed logarithmic.

law, which is written

$$\frac{\Delta I}{I} = \text{constant} = \Delta \,(\text{response}) = 1 \text{ JND} \qquad (8.19)$$

where ΔI is the JND at a level of intensity I. The integral of Eq. 8.19 shows that the response is logarithmic. The fundamental basis for this behavior must lie in the photochemistry of receptor pigments.

So far our discussion has been limited to stimuli which are constant in time and has omitted visual stimuli which vary. A flickering light appears to be blended or fused to seem like steady light above a certain frequency called *critical fusion frequency* (CFF). The CFF depends on the brightness of the light, its wavelength, the stimulus size, and a variety of other parameters. Under ideal conditions and with very bright light, the CFF can be as large as 60 Hz, whereas for very low light levels it can be as small as 5 Hz. The average illumination level for movies and television are chosen carefully so that the frame frequency is higher than the CFF. There is a rather large literature on studies of parameters and their influence on the CFF because it is one of the few ways to measure the human "temporal acuity."

There is virtually no limit to the shortness of a flash light we can detect as long as it is bright enough so that the total light energy delivered to our

eyes exceeds threshold. For short pulses (duration much less than the reciprocal of the CFF) there is a simple reciprocity relationship for detectability
threshold between brightness and duration. For longer pulses we expect that
the average light level would have to exceed threshold. These ideas are very
similar to those presented in Chapter 6 concerning the reciprocity between
pulse length and pulse strength required to initiate an action potential in an
axon.

Finally, no discussion of the limits of our visual system would be complete without mentioning the range of the visible spectrum. We can detect
light of wavelengths between 4000 and 7000 Å, and these limits are imposed
by the combination of a variety of conditions. The transparency of the ocular
media and the range of sensitivity of the photosensitive pigments combine
to provide the primary wavelength limits. In addition, the macula lutea
(yellow pigment) of the fovea tends to limit its sensitivity with respect to that
of the rest of the eye. We are most sensitive to green light with sensitivity
dropping off toward red or blue, as shown in Figure 8.15.

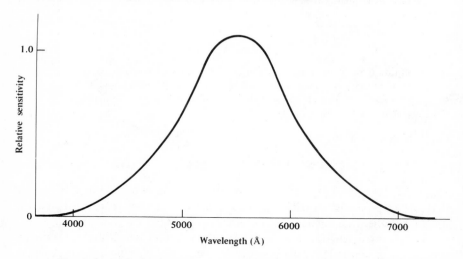

Figure 8.15 The spectral response of the human eye. It peaks in the yellow-green just as sunlight does.

Part 7: Stabilized Images, Acuity, and Illusions

When a person is asked to fix his vision on a particular point, careful observation of his eyes will show that they do not remain fixed. Instead they execute
a combination of different motions called tremor, drift, and saccade. The
tremor motion is a vibration at a frequency about 40 to 80 Hz and of ampli-

tude 1 or 2 minutes of arc. This is superposed on a drift that may last for
$\frac{1}{4}$ to $\frac{1}{2}$ second and cover 5 to 10 minutes of arc. Each drift is terminated by
a saccade or rapid flick that occurs on a millisecond time scale and may also
cover 5 to 10 minutes of arc. Figure 8.16 shows a sketch of typical eye motions
superposed on the retinal cell mosaic.

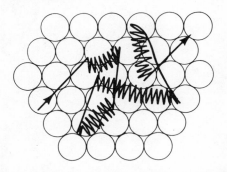

Figure 8.16 Image motion on the retina showing the tremor, drift, and
saccade. The tremor occurs at about 20 Hz, each drift last about $\frac{1}{2}$ sec, and
saccades occur on a millisecond time scale.

It is clear that the image presented to the retinal cells is rapidly moving
around and changing. Any judgments that we make in a visual acuity test
either must be done on a rapidly moving target or must somehow use the
existence of the motion in some sort of signal averaging system.

It is possible to arrange an optical system that causes the image motion on
the retina to be enhanced, reduced, or eliminated completely so that the
retinal image is stabilized. The eyeball is fitted with a special contact lens
having a small flat silvered area near its edge, out of the field of view [see
Figure 8.17(a)]. Then an image is formed using a special projector which
reflects light off the small mirrored area to a screen which is in the field of
view where the subject can see it. When the eyeball and mounted mirror
rotates through an angle θ, the reflected beam is rotated through an angle of
2θ in the field of view. The apparent angle of rotation can be varied using the
system of four plane mirrors shown in Figure 8.17(b) and (c) so that it is also
θ or any other value between 0 and 2θ.

When the mirrors are arranged so that the apparent angle is θ and the
image is therefore stabilized on the retina, the subject reports that the image
seems to disappear. The nature of the disappearance depends on the type of
image (simple versus complex, colored versus monochromatic, etc.) but there
is no question about the fact that it does fade. The conclusion is that our eyes
are basically ac devices and cannot see stabilized images.

Attempts have been made to establish a connection between ocular motion
and acuity. On first thought it seems impossible for humans to detect the
straightness of a line because the eye is not well suited to the job. It has been

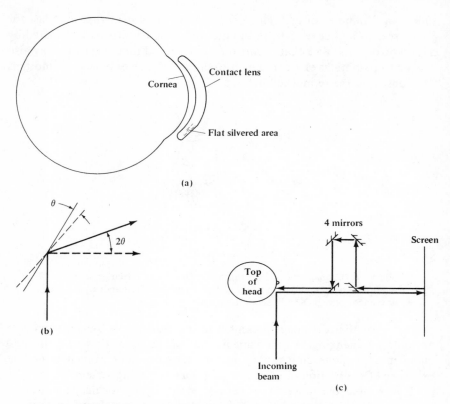

Figure 8.17 Stabilized image measurements. (a) The contact lens on the eye. (b) A ray diagram for the reflections showing that the angle is doubled. (c) The optical system that cancels out the image motion.

proposed that the motion of the eye may help in this problem because only the image of a straight line will be self-congruent after some translational motion. However, the suggestion that we interpret a line as straight only when its image doesn't change under the translation of the scanning motion of the eye does not stand up under close scrutiny. To begin, a wiggly line viewed under the conditions of a stabilized image should appear to be straight. Also, straight lines should soon disappear from view since their image on the retina is not being scanned. Finally, the theory fails to account for many of the common visual illusions shown in Figure 8.18. Each of these illusions show lines that appear to be bent, curved, or broken, but all of them are perfectly straight. Any complete theory of vision dealing with problems of form perception and acuity must be able to explain illusions of this type if it is to be an acceptable theory.

These ideas are also unsatisfactory because they depend on the notion that every cell in the retina is precisely placed and that the brain has detailed knowledge of the location of each of the cells. Since the arrangement of cells

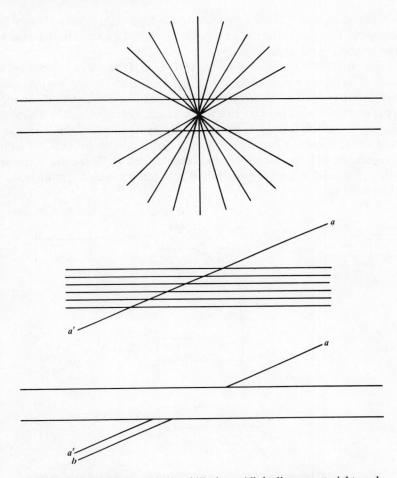

Figure 8.18 Some common visual illusions. All the lines are straight, and the continuous line *aa'* is smooth. In the lower drawing, the line *aa'* is straight.

in the retina is subject to the irregularities of normal growth, and since the cells may very well move around in time, and since some cells may be damaged by cosmic rays or other radiation, no theory that depends on a detailed knowledge of the precise location of retinal cells will be satisfactory.

Part 8: Integrative Processes in the Retina

In an earlier part of this chapter we have seen that the receptor cells are not connected to the brain in a direct and simple manner. We shall now discuss some of the effects of the interconnections between the receptor cells. The two types of synaptic junctions, excitatory and inhibitory, have already been described in Chapter 6. Both kinds exist in the retina and will be denoted here in the symbolism of digital circuits.

Figure 8.19(a) is the symbol of an AND gate. It requires a signal present at both of its inputs simultaneously in order for there to be any output. Figure 8.19(b) is the symbol for an inverter: whenever there is a signal at the input, there is none at the output and vice versa. The combination of these as shown in Figure 8.19(c) functions as an inhibitory synaptic junction. There is no output unless there is a signal present at A and none present at B. Of course, this represents only an output for the receptor cell A. There must also be a parallel channel for the receptor cell B and the combined circuit is shown in Figure 8.19(d).

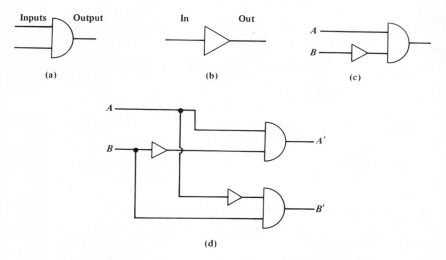

Figure 8.19 Logic diagrams. (a) An AND gate. (b) An inverter. (c) An inhibitory synaptic junction with 100% inhibition. (d) The wiring diagram for lateral inhibition.

An arrangement of synaptic junctions that functions like the circuit of Figure 8.19(d) is said to exhibit *lateral inhibition*. Imagine two adjacent receptor cells in the retina with lateral inhibitory connections between them. If either of the cells is stimulated, there will be a response. If both cells are stimulated, however, the response from each of them will be strongly inhibited because of the excitation of the other cell. Figure 8.20 shows the results of such an experiment in receptor cells from the eye of the horseshoe crab, *Limulus* (often used for visual studies because of its neurophysiological convenience and simplicity). It is clear that the effects of lateral inhibition are present here at the single cell level. We shall soon see their macroscopic importance.

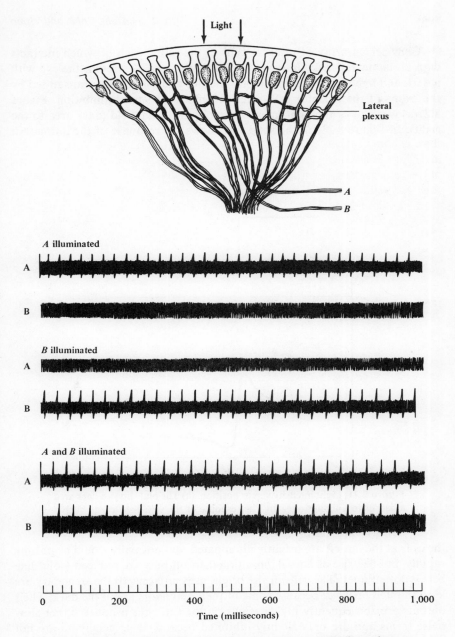

Figure 8.20 Mutual lateral inhibition recorded electrically from the eye of the horseshoe crab. *A* and *B* alone produce larger responses than *A* and *B* together because of the mutual lateral inhibition.

Consider an array of receptor cells in the retina each of which interacts with all its nearest neighbors via lateral inhibition. We present the eye with a dark and light area adjacent to one another and ask what is transmitted by the axons of the receptor cells after the effects of lateral inhibition. Figure 8.21(a) is a plot of the light intensity across the visual field (dark area to the right). In Figure 8.21(b), the solid line shows the response of the retina. All

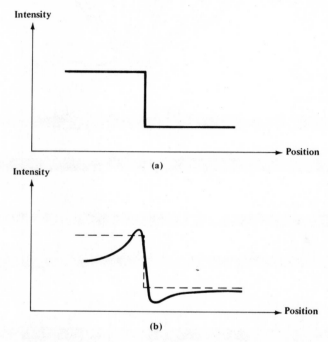

Figure 8.21 Light intensity versus position. (a) The plot shows a dark area adjacent to a lighter one. (b) The response of the visual system to such a stimulus as a result of lateral inhibition. Notice the edge enhancement.

the cells at the far left are brightly illuminated and normally would be pulsing rapidly, but because of lateral inhibition their outputs are reduced (solid line below dotted line). The cells in the bright area adjacent to the boundary are less inhibited however because some of their neighbors are in the dark. Their outputs are consequently higher. Cells on the far right produce rather slow pulse trains and are only slightly inhibited because their neighbors are not very strongly excited, but those in the dark near the boundary are strongly inhibited by their highly excited neighbors and their output is lowered. As the solid line indicates, lateral inhibition serves as an edge enhancement mechanism for the eye.

When a distinct edge is brought into the field of view as above, we might expect (from Figure 8.21) that the edge would show a dark band just inside the darker area and a light band just inside the light area. These bands, called Mach bands, are easily observed in the shadow of a straight edge in a normally lighted room (see Figure 8.22). The experimental arrangement of Figure

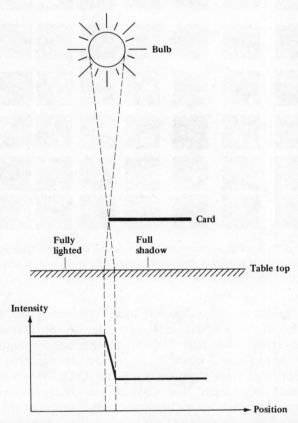

Figure 8.22 Illustration of how to observe Mach bands in a well-lighted office or classroom. It's best if the edge of the card is parallel to the fluorescent light tubes.

8.22 works best with fluorescent lights when the edge and its shadow are arranged to be parallel to the straight fluorescent tubes. The Mach bands are easily observable in this striking demonstration. The appeance of gray areas at the intersections of the white strips in Figure 8.23 is also readily explained in terms of Mach bands.

In Part 7.5 there is a detailed discussion of how lateral inhibitory mechanisms can enhance constrast and sharpen edges. It is easy to see how inhibi-

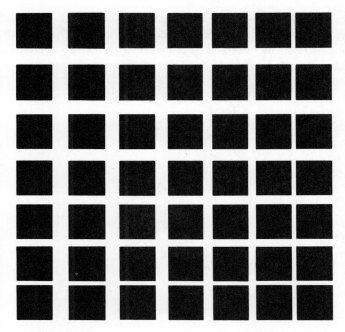

Figure 8.23 The famous pattern of disappearing gray spots. Because lateral inhibition is much less common in the fovea, looking at a gray spot causes it to vanish.

tory sharpening can partially offset the effects of blurring and other image degradation by the optics of the eye. In particular we note that visual acuity is undoubtedly improved by lateral inhibition. Appendix I contains a discussion of the effects of lateral inhibition in art and photography.

The perception of edges plays a very important role in the visual process. Edges determine outlines and shapes, which are usually the first thing we seek in an image or draw when we begin a sketch. The ability of a cartoonist to caricature an individual with just a few lines provides further proof of the importance of edges.

In 1971, Edwin Land and John McCann published the results of an experiment that was explained with the hypothesis that edges are primarily responsible for our perception of lightness and darkness. The experimental subjects were asked to view a collage of various gray squares and rectangles of different brightnesses. When viewed under uniform illumination, a rank order of brightnesses could be readily established for the various areas. When viewed under very nonuniform illumination, the *same* rank ordering of brightness prevailed. A light area in very dim illumination looked lighter than a dark one in very bright illumination even though there was far more light reaching the eye from the brightly lit, dark area! Both photometric measurements and isolated viewing of the two areas confirmed that there was more light coming to the eye from the area that appeared darker. Land and McCann concluded that our perception of lightness depends on the contours and edges and not on the intensities. They proposed the existence of retinex elements (retina and visual cortex) that act in some way to process the data from the visual field so that we are aware of the nonuniform illumination and account for it in our judgments. The published article even contains sample slides that the reader can view and use to perform his own experiments (see Appendix I for a different example).

The lateral inhibition between cells does not have to exist between nearest neighbors only. In order to extend our ideas about retinal information processing, we need to consider another element of digital logic circuitry. Figure 8.24(a) shows the symbol for an OR gate. It has an output if there is a signal

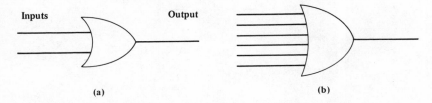

Figure 8.24 OR gate with two (a) and many (b) inputs.

at either one or both of its inputs. An OR gate can have several inputs, as shown in Figure 8.24(b), and there is an output signal when a signal is present on at least one of the inputs. The OR gate acts as a summing junction in a way very similar to an excitatory synaptic junction.

Consider the array of receptor cells in Figure 8.25(a). We imagine that the outputs of all the cells in the central spot are connected to a single OR gate as shown and that the outputs of all the cells in the surrounding annulus are connected to another OR gate. The output of one OR gate is inverted and connected, along with the output of the other one, to the inputs of an AND gate. There is output from the AND gate only when the visual field corresponding to the chosen cells is presented with a bright spot on a dark surround. If the area is uniformly illuminated, there will be no output. Axons that respond only when a bright area with a dark surround is present in a particular part of the visual field have been found in cats and other mammals, and the visual fields are called *on-centered regions*.

Of course, the output of the OR gate from the central region (instead at from the periphery) could be inverted before going to the AND gate making an off-centered region from the same cells. Such circuits are not mutually exclusive, as shown in Figure 8.25(b). The idea of multiple use of the same

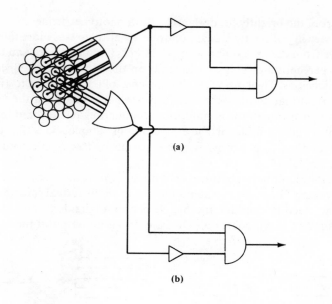

(a)

(b)

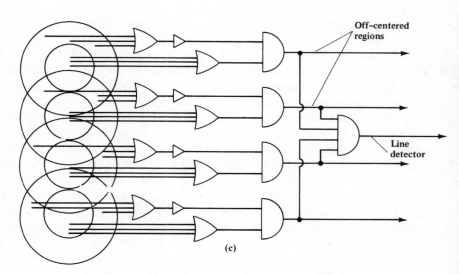

(c)

Figure 8.25 Circuit diagram of on-centered region (a). For an off-centered region one simply inverts the output from the central and peripheral regions and sends them to another gate (b). In (c) the combination of several spot detectors in a row produces a line detector.

cells can be extended considerably, as shown in Figure 8.25(c). Here the cells in the top annulus are connected to the top OR gate, while those in the second annulus are connected to the next OR gate and so on. Since some cells are found in more than one annulus, they are connected to several OR gates. The same is true for the cells in the central region of each of the overlapping on-centered regions. Some cells are found in the annulus of one visual field and in the center of another. The entire array of cells is arranged as a contiguous row of off-centered regions. We next imagine an AND gate with many inputs instead of just two that requires that all the inputs have a signal in order to produce an output. If we bring all outputs of the off-centered regions to the inputs of this multiple AND gate, we have constructed a line detector. Its output carries a very special message to the brain: There is a dark line on a light background oriented at a particular angle. Such line detectors have been observed in the visual system of the cat. The result of presenting lines at various angles to the visual field corresponding to a line detector are shown in Figure 8.26. Unless the orientation is correct, there is no response.

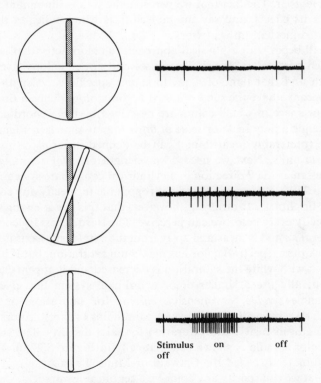

Figure 8.26 The response of line detector in a cat to several lines at various orientations. Only the vertical line elicits significant response.

We have seen that information sent from the eye to the brain is not a point-by-point description of the light intensity in the visual field. It has been processed, interpreted, and perhaps specialized for some purpose (e.g., an off-centered region is an ideal bug detector for a fly-eating frog). This peripheral data processing may very well be called brain work and leads some people to claim that the retina is very much a part of the brain. There is certainly adequate support for that claim.

Part 9:   Colorimetry

Before we begin the study of the problem of color perception, it is necessary to study color specifications at some length. Even a casual observation is sufficient to inform us that color perception is a complex phenomenon. For example, a white tablecloth appears white when it is illuminated only by a yellow candle flame and therefore is reflecting light to the eye that is as yellow as the flame itself. The fact that we perceive the yellow illuminant, somehow correct for its effects, and say the tablecloth is white indicates that a very sophisticated mechanism is at work.

We shall begin with a physical approach to colorimetry that is based on well-defined parameters. We shall soon see that not all colors have a specific wavelength and that light of a predominant specific wavelength does not always appear to have the same color. It is common to define a three-dimensional color space in which colors are described by their coordinates. First we shall define a gray scale or *scale of intensities* as shown in Figure 8.27(a). This scale (preferably logarithmic) can be defined in terms of any suitable photometric units. Next we need a wavelength or *color coordinate* that is usually chosen as the θ direction in a cylindrical coordinate system, as shown in Figure 8.27(b). By varying θ and z it is possible to specify any color at any particular brightness. It turns out, however, that this is not enough information to specify every color we can perceive. The third necessary coordinate is called *saturation* and is measured away from the z (intensity) axis in the radial direction. A pure (spectral) color has maximum saturation, but if it is diluted by mixing it with white the saturation is decreased and it appears as a pastel. Artists generally use a similar color coordinate system but give the axes different names: *value* for intensity, *chroma* for saturation, and *hue* for wavelength. We have all seen arrays of paint chips or fabric colors arranged by wavelength, intensity, and saturation for sales displays. The entire color space or color spindle is shown schematically in Figure 8.27(c). This set of color coordinates is called the Ostwald or Munsell system.

Now that we can specify any color by a set of three numbers, we ask what happens when we mix two colors. Is there a way to find the coordinates of

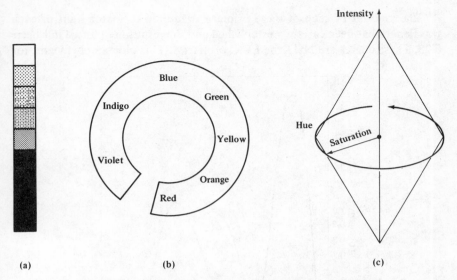

Figure 8.27 (a) The gray scale. (b) The circle of hue. (c) The color coordinate system of the Ostwald description of colors.

the new color that results from the mixing if we know the coordinates of the original colors? The answer is, of course, yes, but the rules for doing it in this system are very complicated. In order to establish a straightforward set of rules, we shall investigate a different system of color coordinates.

We shall begin by defining primary colors. It has long been known that most color sensations can be reproduced by mixing together other colors. There have been very many color matching experiments in which subjects are asked to vary the relative intensities of the three colors to be mixed (called primaries) until they produce a satisfactory match to the original sample. The principle conclusions from these experiments are that the relative mixture is very well defined and repeatable for a particular sample and that no more than three primaries are needed for the match. There are a great many colors that can be produced with only two primaries, but it always takes three to span the entire range of color sensation. Of course, the amounts of the primaries needed to match a particular color is different when a different set of primaries is chosen. We shall choose a specific set of primaries called the I.C.I. (International Commission on Illumination) or C.I.E. (*Commission Internationale de l'Eclairage*) standard observer primaries. These do not correspond to any real colors but represent an artificial construct from which it is convenient to define other colors. The reader should recognize that the choice of a set of primary colors has no more significance than the choice of any other coordinate system: sometimes the proper choice of a coordinate system makes a particular problem easier to solve.

The amount of each of these primaries required to match light of each wavelength has been carefully established and the results are plotted in Figure 8.28. These curves are NOT the I.C.I. primaries! This plot simply means, for

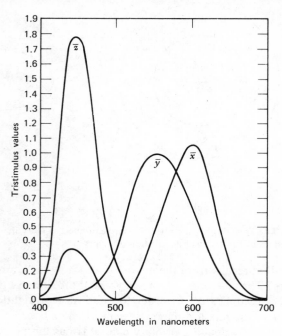

Figure 8.28 Tristimulus values for the "standard observer" agreed upon by C.I.E. in 1931. The *y* curve has been chosen to match the photopic (daylight) luminosity curve of the normal eye.

example, that to match the sensation of 5000 Å requires approximately equal amounts of the I.C.I. primaries $\bar{y}$ and $\bar{z}$ along with no component of $\bar{x}$. The amount of the I.C.I. primaries required to match any particular color, whether it is a pure spectral color or not, is customarily denoted by a capital letter (X, Y, or Z) whereas the names of each of the primaries (the basis vectors or coordinate axes) are denoted $\bar{x}$, $\bar{y}$, $\bar{z}$. Therefore the I.C.I. formula for a particular color A is given by

$$A = X\bar{x} + Y\bar{y} + Z\bar{z}. \tag{8.20}$$

We define a new set of coefficients called *chromaticity coordinates* for any particular color

$$x = \frac{X}{S}, \quad y = \frac{Y}{S}, \quad z = \frac{Z}{S}, \quad S = X + Y + Z. \tag{8.21}$$

Since $x + y + z = 1$, it is clear that only two chromaticity coordinates are needed to specify a color, and a two-dimensional plot called a *chromaticity diagram* can be made (see Figure 8.29). The curved boundary represents the locus of the spectral colors (the coordinates can be taken from Figure 8.28) and the region in the middle represents somewhat less saturated colors.

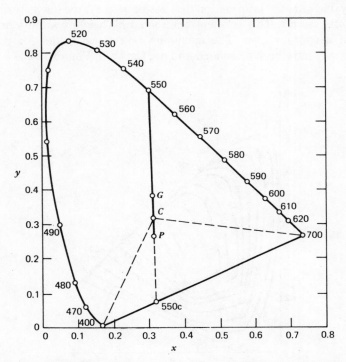

Figure 8.29 The chromaticity diagram. The sample *G* lies on the line between the illuminant *C* (daylight) and wavelength 5500 Å. This identifies it as a partially saturated yellow-green.

The chromaticity diagram has a variety of interesting properties that can be found from Eqs. 8.20 and 8.21. A region in the center appears white, but each particular point in the region corresponds to a particular white (carbon arc, daylight, etc.). The I.C.I. standard white illuminant *C* is shown in Figure 8.29. A line drawn from this point through the point that represents any particular color intersects the boundary at the perceived dominant wavelength of that color in that particular light. If the illuminant is different, one simply starts from a different point. Furthermore, if two colors whose coordinates are known are mixed together, the resulting color will lie on a line joining their two points at a distance determined by the proportion of each of the colors. *Complementary colors* are those that, when mixed, give a neutral gray or white. It is clear that complementary colors are those whose locations on the chromaticity diagram can be joined by a straight line passing through some "white" or neutral area. Several other properties of the chromaticity diagram can also be determined from Eqs. 8.20 and 8.21.

In collapsing the three variables of color space into two to make the chromaticity diagram, the information that has been lost is the intensity or brightness of the sample. This is often represented by height contours much like those of a contour map. The maximum reflectance values for particular chromaticity have been calculated and are shown in Figure 8.30.

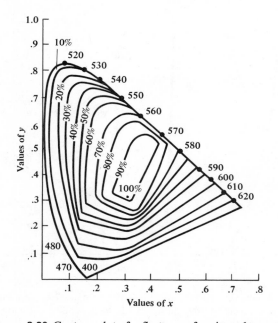

Figure 8.30 Contour plot of reflectance of various chromaticities.

We have discussed here the results of color mixing as if we were adding the light from two different colors together to form a third one. This process, called *additive color mixing*, is to be contrasted with *subtractive color mixing*. If white light is passed through a filter that removes some of the light leaving a colored beam and that beam is then passed through a different filter, the new beam has a color defined by a subtractive color process. Subtractive color mixing is based on different primaries and different rules than those of additive color mixing.

An understanding of the chromaticity diagram, the necessity for three primary colors, and the rules of color mixing has prepared the way for a study of theories of color vision.

Part 10 : Color Vision

It is believed that color vision occurs only in the fovea because color discrimination is very low in the peripheral region of vision even though luminous sensitivity is very high. Also color discrimination, like visual acuity, is very high only in the central part of the field of view corresponding to foveal vision. For this reason it is believed that rod cells do not discriminate colors and that color vision occurs only in the fovea where there are cones.

Color vision deficiencies in certain people can be used to provide information about color vision. These defects can best be categorized by reference to the chromaticity diagram. Anytime an attempt is made at color matching, there is always some range of uncertainty in the amount of each primary to be mixed with the others. This range of uncertainty defines a region of finite size on the chromaticity diagram, instead of a precisely defined point, over which a subject reports a color to be matched. People with "normal" color vision have well-defined error regions on the chromaticity diagram that are not all necessarily the same size. There are some people, however, who have abnormally large regions of uncertainty in certain areas of the chromaticity diagram. If the abnormally large region occurs in the red part of the chromaticity diagram, we say the person is "red blind" and call him a *protanope*; if it is in the green region, he is called a *deuteranope*. There are some people with multiple error regions so large that they overlap one another, and we say these people cannot differentiate between completely different colors. There are even some people whose vision is completely monochromatic. Color blindness is a hereditary, sex-linked characteristic.

The rules of color mixing and the existence of exactly three primaries has led to certain theories of color vision that are supported by the data on color blindness and the idea that color vision occurs only in the cones. In 1802 Thomas Young proposed that human color vision is mediated through three types of fibers or cells in the optic nerve and that the rules of color mixing applied to these three types of receptors. These cells were assumed to have their sensitivity to different wavelengths vary across the spectrum and correspond to the red, green, and blue primaries, as shown in Figure 8.31. This idea was extended by Helmholtz and has formed the basis for the most widespread and useful theory of color vision.

Physiologists have searched for evidence that there are either three different kinds of cone cells or that there are three different photopigments to be found in the cones. The experiments are extremely difficult because they must be done in single cells and because the visual pigments deteriorate very rapidly after the removal of the eye from the experimental animal. Only in the 1960s has there been convincing evidence for the three different cone cells. A plot of the different spectral absorption for goldfish retinal cells is shown in Figure 8.32.

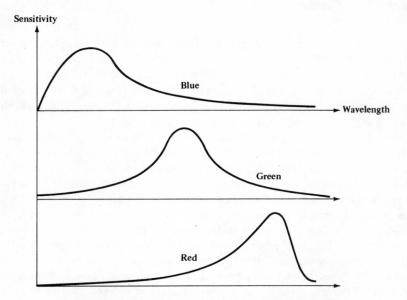

Figure 8.31 The red, green, and blue pigments of Helmholtz.

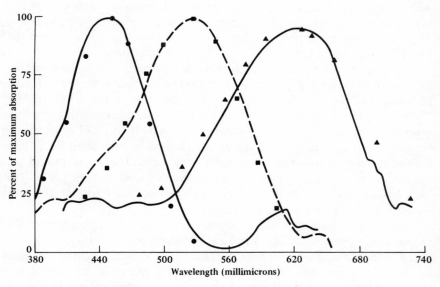

Figure 8.32 Spectral characteristics of goldfish cone pigments showing three, well defined types.

The basis of the *three-color theory* has now been established. Its fundamental notions are the same as those proposed by Young and Helmholtz, although supported by new and better data than was available to them. For example, it is found that people with certain color defects can achieve satisfactory color matching with only two primaries instead of the usual three, and it is proposed that these people are missing one of the three types of cones.

In the course of testing the three-color theory, a variety of observations has led to alternative theories of color vision. The most notable of these is the *opponent theory* of Ewald Hering. He suggested that the color vision mechanism, although tripartite as in the three-color theory, is based on three separate systems that are each composed of complementary or opponent processes. One of these is simply light-dark discriminator, the second is a red-green detector, and the third is a blue-yellow detector. Hering's theory has the complementary color ideas built into it, and the unique notion that there are four "primaries" rather than the traditional three.

The theory is supported by a number of phenomena including simultaneous color contrast and the phenomenon of negative afterimage. *Simultaneous color contrast* is observed in a variety of circumstances where a single color is viewed in the presence of different backgrounds. It has long been known by artists that the perception of a color is strongly influenced by the presence of other colored areas surrounding it. Hering's theory can be used to predict the influence of various surroundings on colors on the basis of competition between the opponent colors (red-green and blue-yellow). In the *negative afterimage* experiment a subject views a brightly illuminated field of a single color or an array of colors. When the field is abruptly shut off and replaced by either a dark or a neutral field, the subject reports the presence of the original objects appearing in their complementary colors. If the stimulus is replaced by its complement instead of a neutral field, the sensation is enhanced and the subject reports brilliantly saturated colors. A favorite demonstration of this phenomenon is the projection of a 35-mm slide of an American flag with green and black stripes and black stars on a yellow field. When the slide projector is turned off, the red, white, and blue flag appears in all its glory. The phenomenon is explained in terms of adaption and/or fatigue. *Retinal fatigue* occurs under a constant stimulus and is observed as an apparent reduction in saturation or intensity after some period of time. When the stimulus is removed, the adapted receptors overreact and, in the case of the Hering model, produce the apparent sensation of the complementary or opponent color.

Many artists paint using small dabs of different colors rather than mixing the colors on their palettes. They depend on color mixing in the eye by virtue of the fact that the individual patches of color are too small to be resolved at any reasonable viewing distance. Of course, the color mixing process is then additive rather than subtractive, which generally results in brighter colors and higher saturation. Some impressionist painters found that an effect of sparkling brilliance can be achieved when the patches of color are large enough to be resolved separately, and the following explanation is proposed. This sparkling effect is most noticeable when the patches of adjacent color are complementary, and the effect is attributed to a negative afterimage enhancement brought about by eye movements of the type described in Part 8.7. After a small group of receptors has become fatigued by detection of a particular color, a saccade moves that color off those receptors and replaces it with the complement that is found in an adjacent color patch. The result is an enhanced impression of the colored patch and the sparkling effect. It is very noticeable in some of Monet's paintings where there is sunlight on water (e.g., "The Houses of Parliament at Sunset," National Gallery of Art, Washington, D.C.).

Many theories of color vision have been proposed and there have been books written on the subject. Instead of a further discussion we shall describe a series of simple and beautiful experiments in color vision that appear to contradict the three-color theory. These experiments were performed by E. H. Land of the Polaroid Corporation in the late 1950s and early 1960s.

Land made two black-and-white photographs of a typical scene using a red filter on the camera for one picture and a green filter for the second one. The photographs, mounted as slides, were black and white because of the film he used. The two slides were placed in separate projectors and the images were superimposed on a screen. The view was, of course, still black and white. Next a red filter was placed over the lens of the projector carrying the slide originally exposed through a red filter (called *long record* in reference to the filter wavelength). Instead of seeing the expected pink, red, and gray image predicted by our established color mixing laws, the entire scene appeared in all of its original colors. This fantastic demonstration seems to be totally incompatible with all known theories of color vision. There have been many attempts to explain the "Land effect" in terms of traditional theories, but there has been only limited success.

Edwin Land and John McCann have extended their retinex theories (discussed in Part 8.8) to include color vision. They have carefully constructed a theory which is consistent with the anatomical existence of three different color pigments and psychophysical data of three primary color matching, while including a mechanism which describes the two color experiments. The theory is based on contrast detection at edges and is supported by a series of careful experiments.

In one of these experiments, subjects view an arrangement of squares and rectangles of various colors. A special telescopic photometer is used to determine the color coordinates of some particular area (say green) of the image under some particular illumination. The color of the illumination (provided by three independent projectors) is now changed so that the color coordinates of the light reflected from some other area of the collage (say yellow) are exactly those that had previously elicited the sensation of green. The area so illuminated still appeared yellow! Furthermore, all the areas of the collage retained their previous colors, even though the color of the illumination was strong enough to make a green paper reflect light whose chromaticity coordinates were those of yellow. Similar effects, called *color constancy*, have been known for many years, but these experiments are more careful and quantitative than earlier ones. Land concluded that our perception of color depends on relative lightness, which, as we have seen in his earlier experiments, depends on contours and edges.

Land wrote "the mystery is how we can all agree with precision on the colors we see when there is no obvious physical quantity at a point that will enable us to specify the color of an object." In order to solve that mystery Land and McCann have proposed the retinex theory of color vision with four separate classes of retinex elements. One class contains only the monochromatically sensitive rod cells, and the other three classes are each made up of the different cone cells: those sensitive to long, medium, and short wavelengths. Each of these classes of retinex elements determines the relative lightness of various areas in an image, and these relative lightnesses are compared with each other. The result of these comparisons produces the sensations of color.

Since the theory includes the three-primary system of the different cones, it is consistent with various tests of three-color theory. Since it is independent of the detailed nature of receptor cell response, it is consistent with simultaneous color contrast and negative afterimage data. Since much color information can be gained from the excitation of only two retinex systems, the theory explains Land's two color experiments. Finally, because the retinex elements take into account illumination anomalies, the theory provides a satisfactory explanation of why the tablecloth illuminated by yellow candlelight looks white.

The retinex theory of color vision will have to withstand extensive tests and measurements. Whether it will survive or not depends on how well it encompasses all the data from decades of testing other color vision theories. If it fails for some reason, any new theory will have to explain all the older data as well as the experiments performed by Land and McCann in the course of developing this theory. At present, we still don't know how we see colors.

APPENDIX I
LATERAL INHIBITION AND PAINTING

The process of recording an image is much more difficult and complex than that of recording a sound. Whereas a sound is simply a time-dependent pressure variation (one dimensional), an image must be recorded in two dimensions and must display color variations as well as intensity variations. In this discussion we shall be concerned only with monochromatic (black-and-white) images.

An outdoor scene in natural sunlight typically has some very bright areas, some very dark areas, and a nearly continuous range of brightnesses between the extremes. It is quite possible for the extremes to vary by a factor of $10^5 : 1$ or $10^6 : 1$ in brightness. Even an indoor scene may have a variation of $10^3 : 1$ to $10^4 : 1$ in brightness. On the other hand, the usual methods of reproducing images are very limited in their range of brightnesses: photographs are limited to a range of about $15 : 1$ and oil on canvas has a range of slightly less than $10 : 1$. The two questions that come to mind immediately are "How is the intensity scale compressed when either of these processes is used to record a scene?" and "Why do paintings generally look much more realistic than photographs?"

The relationship of relative intensities in a scene and an image (or between two images) is described by a transfer function. Since the brightness often ranges over several orders of magnitude, it is customary to plot scene brightness versus image brightness on a log-log graph and call the result the *characteristic curve*. Figure I.1(a) shows a typical curve for the photographic process. Different characteristic curves can be obtained by using different films, chemicals, temperatures, and processes in the production of a photograph.

The linearity and constancy of the photographic characteristic curve has been studied and perfected over the years in the science of photogrammetry. The curves are reliable, repeatable, predictable, and uniformly controllable. In fact, it is precisely these properties that are responsible for the flat and somewhat unrealistic appearance of photographs. By contrast, a painting can look so real that we are often fooled, even if only for a moment, into believing that the subject of a painting is a real scene and not simply a picture.

The brightness transfer function of a painting is completely under the control of the artist. He can make its slope flat or even negative in places, he can vary it from one part of the painting to another, and he can make discontinuous breaks in it. The fact that it can be varied in different regions of the same painting makes it undefinable in any sense as simple as the photographic characteristic curve, but examples of some kinds of variations are shown in Figure I.1(b). Certain characteristics of the transfer function might be said to define some aspects of the style of one artist's work that serves to

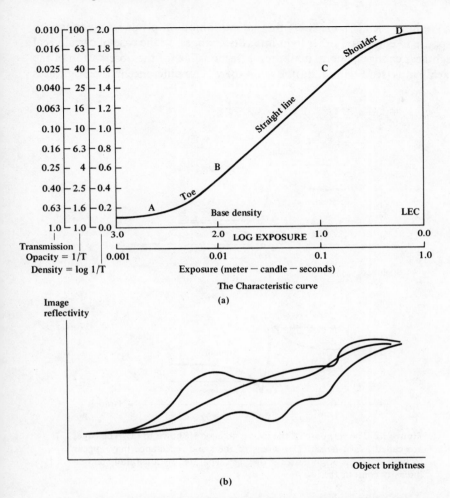

The Characteristic curve

(a)

(b)

Figure I.1 A typical characteristic curve for photographic films is shown in (a). Possible curves an artist may use are shown in (b). The artist is not confined to regular curves or to the same curve in every part of a given painting.

distinguish it from that of a different artist. In one sense, these variations from one artist to another constitute one of the elements of artistic style or technique.

One of the artist's principal tasks is to convey the impression of a much wider range of brightness than the oil-on-canvas medium allows. This requires the production of a visual brightness illusion that must be carefully concealed from the viewer. The artist takes advantage of lateral inhibition, Mach bands, and related phenomena to accomplish this illusion.

We have seen (Figure 8.21) that lateral inhibition produces an enhanced response to edges and other light intensity changes. In the case of certain real brightness changes, lateral inhibition is responsible for the observation of the Mach bands [see Figure I.2(a)]. Even if there is no difference in the brightness

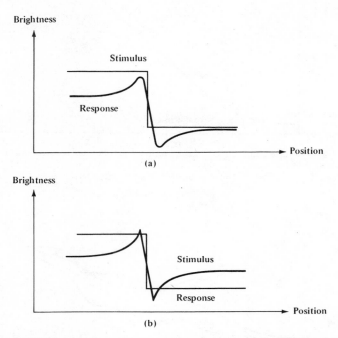

Figure I.2 The response of the eye to an edge is shown in (a). The edges are called *Mach bands*. Two areas of the same reflectance may appear to have different brightnesses if they are separated by a carefully chosen intensity contour (b).

on either side of a boundary, the presence of certain boundaries can produce the illusion of a brightness difference. For example, the plots in Figure I.2(b) illustrate the Craik-O'Brien contour and the response to it. In their 1971 paper, Land and McCann published a striking demonstration (reproduced in Figure I.3) showing the effects of boundaries on lightness perception. The two square areas appear to have very different brightnesses, but it is easy to demonstrate that they are the same by laying a pencil vertically on the boundary between them.

A white disc with a dark sector as shown in Figure I.4 can be used in order to demonstrate further the effects of contours on brightness perception. When the disc spins, the inner and outer regions have the same brightness, but the existence of the contour between them makes the inner region appear

Figure I.3 An example of the phenomenon of contour dependence. Place a pencil along the vertical boundary and the apparent brightness difference between the two halves of the image vanishes.

Figure I.4 A Craik-O'Brien contour can be produced by spinning this disk. The reader may experiment with a variety of contours, double contours, reverse contours, etc., with carboard, ink, and an electric drill.

brighter. Reversing the contour makes the inner region seem darker. An appropriately chosen contour between two regions can even make one that is actually darker appear to be brighter. The reader is urged to see **Ratliff's** 1972 *Scientific American* article and his book for a discussion of similar phenomena in Oriental pottery as well as in other image production processes (such as xerography). Artists take advantage of these effects in order to create the illusion of greater or less brightness in areas of a painting where no such differences really exist. In this way they can paint a picture that has very large apparent brightness differences even though the medium is restricted to a considerably smaller intensity range. Because they can vary the use of this technique in different parts of a picture, they can create a much more "realistic" looking image. This is especially true in charcoal or pencil drawings. The technique, called *chiaroscuro*, is found in a wide range of paintings. Rembrandt was a master of these illusions (Rembrandt lighting). The neo-impressionist Seurat actually exaggerated the contours to create additional lighting effects.

We have seen that it is possible to use appropriately chosen contours to change the perception of brightness. The illusions produced by artists give paintings a brilliance and vividness that is not possible to achieve in photographs. The difference exists because artists can control the brightness transfer function, vary it from one part of an image to another, and augment it with appropriately chosen color contrasts. The uniformity of the photographic process precludes this artistic touch and therefore restricts the illusion of reality.

Questions and Problems for Chapter 8

1. Discuss some of the characteristics of an eye that could see electromagnetic radiation whose frequency varied over a factor of 1000 instead of a factor of 2. Discuss for radiation of both shorter and longer wavelengths than visible.

2. Why is the index of refraction of a medium always larger than unity? What kind of medium would have an index less than one?

3. Suppose that a beam of light traveled from a medium of greater index of refraction to a medium of lesser index at an angle such that $\sin \theta_2 = (n_1/n_2) \sin \theta_1 > 1$. What is θ_2? What happens?

4. The cornea of the eye has a radius of curvature of 7.7 mm, but the radius of the eye is about 12 mm. How can that be?

5. Draw a schematic diagram of the feedback control system that regulates focusing of an image by the eye. Indicate the sensor, amplifier, and effector.

6. Binoculars are characterized by two numbers: the first is the magnification and the second is the diameter of the objective. Common binoculars are 7×50 and there are only small variations from these specifications. Why aren't 12×50 binoculars common?

7. Show that the optical system of Fig. 8.17 can provide an angular deviation of anywhere from 0 to 2θ by proper choice of distances.

8. Discuss the fading of stabilized images on the retina in terms of nerve fatigue.

9. What aspects of our intensity perception must be emphasized in order to explain the experiments of Land and McCann?

10. It has been said that the retina is a part of the brain. What is the evidence for such a claim?

11. Why are three primary colors chosen? Why not two or four?

12. Distinguish between the color variables $\bar{x}\,\bar{y}\,\bar{z}$, $x\,y\,z$, and $X\,Y\,Z$. Why is $x + y + z = 1$?

13. Use Fig. 8.28 and graphically determine a few of the points on the periphery of the chromaticity diagram, Fig. 8.29.

14. Show that the coordinates on the chromaticity diagram for a mixture of colors can be determined by simple algebra. Use Eq. 8.20 and 8.21 to determine the coordinates of some mixtures.

Bibliography

General

1. G. BEGBIE, *Seeing and The Eye* (Natural History Press, Garden City, N.Y., 1969).

2. F. A. GELDARD, *The Human Senses* (John Wiley and Sons, New York, N.Y., 1972).

3. R. L. GREGORY, *Eye and Brain* (McGraw-Hill Book Co., New York, N.Y., 1966).

4. M. MINNAERT, *Light and Color* (Dover Publications, Inc., New York, N.Y., 1954).

5. C. RAINWATER, *Light and Color* (Golden Press, New York, N.Y. 1971). (An excellent, authoritative, low cost book.)

Special References by Section

Section 1 and Section 2

1. D. HALLIDAY and R. RESNICK, *Physics* (John Wiley and Sons, New York, N.Y., 1968). (Or any comparable introductory physics text.)

Section 3

1. E. ACKERMAN, *Biophysical Science* (Prentice-Hall, Inc., Englewood Cliffs, N.J., 1962). See new edition, Ref. 2 of Ch. 6.

2. T. RUCH and J. FULTON, *Medical Physiology and Biophysics* (W. B. Saunders Co., Philadelphia, Pa., 1960). (An encyclopedia text.)

Section 4

1. E. ACKERMAN, *Biophysical Science* (Prentice-Hall, Inc., Englewood Cliffs, N.J., 1962). See new edition, Ref. 2 of Ch. 6.

2. B. JULEZ, "Texture and Visual Perception," *Scientific American* (February 1965).

3. T. RUCH and J. FULTON, *Medical Physiology and Biophysics* (W. B. Saunders Co., Philadelphia, Pa., 1960).

Section 5

1. H. METCALF, "The Stiles-Crawford Apodization," *J.O.S.A.* **55**, 72 (1965).

Section 7

1. R. GREGORY, "Visual Illusions," *Scientific American* (November 1968).

2. M. LUCKIESH, *Visual Illusions* (Dover Publications, Inc., New York, N.Y., 1965).

3. J. R. PLATT, "How We See Straight Lines," *Scientific American* p. 121. (June 1960). (The theories proposed here are clearly inadequate.)

4. R. PRITCHARD, "Stabilized Images on the Retina," *Scientific American* (June 1961).

5. L. RIGGS and ULKER-TULUNAY, *J. Opt. Soc. Am.* **49**, 741 (1959).

Section 8

1. D. HUBEL, "The Visual Cortex of the Brain," *Scientific American* (November 1963).

2. E. LAND and J. MCCANN, *J. Opt. Sci. Am.* **61**, 1 (1971). (Description of retinex theory.)

3. C. MICHAEL, "Retinal Processing of Visual Images," *Scientific American* (May 1969).

4. W. MILLER, *et al.*, "How Cells Receive Stimuli," *Scientific American* (September 1961).

5. F. RATLIFF, "Contour and Contrast," *Scientific American* (June 1972).

6. ———, ed., *Studies on Excitation and Inhibition in the Retina* (Rockefeller University Press, New York, N.Y., 1974).

7. *Image, Object, and Illusion* (W. H. Freeman and Co., San Francisco, Calif., 1974). (Book of readings from *Scientific American*.)

Section 9

1. Encyclopedia Britannica article on *Color*.

Section 10

1. E. LAND, "Experiments in Color Vision," *Scientific American* (May 1959).

2. E. LAND, "The Retinex Theory of Color Vision," *Scientific American* (December 1977).

3. E. MACNICHOL, "Three Pigment Color Vision," *Scientific American* (December 1964).

4. R. TEEVAN and R. BIRNEY, ed., *Color Vision* (D. Van Nostrand, Princeton, N.J. 1961).

9

EXPERIMENTAL
TECHNIQUES

Part 1: Optical Spectroscopy

A. In Chapter 8 we discussed the properties of electromagnetic radiation with emphasis on visible light. The domain of optical spectroscopy spans a considerably larger range of frequencies, much of which is not visible. The shortest wavelengths form the vacuum ultraviolet region that is so named because air absorbs most of the light and spectroscopy in this region must be done completely in vacuum. It spans the wavelength range from about 200 to about 1800 Å. For wavelengths longer than about 1050 Å, LiF crystals are transparent and can be used for windows. For wavelengths longer than about 1200 Å the more stable material MgF_2 is transparent as well. For the shorter wavelength region of the vacuum ultraviolet (sometimes called hard ultraviolet) there are no good materials for windows or lenses.

The wavelength region from the edge of the vacuum ultraviolet at 1800 Å to the threshold of visibility at about 4000 Å is called ultraviolet. It is generally advisable to use quartz optics since most glasses and plastics have poor transmission. The visible region occupies the wavelength region from 4000 to 7000 Å and the infrared spans slightly more than the next decade of wavelengths up to about 100,000 Å (usually written 10 μ where μ stands for micron, 10^{-6} meter).

It is the job of spectroscopic instrumentation to measure the intensity of light from a source at different wavelengths in order to determine the spectrum (see Chapter 7 for a discussion of acoustical spectra). The most common type of spectrometer works by collecting the light to be analyzed and focusing it into a beam, sending it through some kind of wavelength dispersing element, and analyzing the outgoing light. A schematic diagram of a simple spectrometer is shown in Figure 9.1(a).

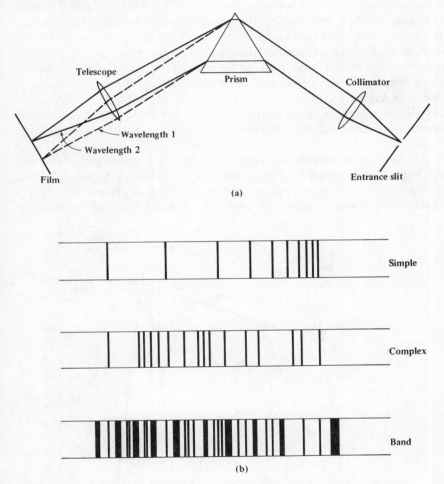

(a)

(b)

Figure 9.1 (a) The ray-optical diagram of a prism spectrometer used photographically. (b) Spectra of increasing complexity.

Light entering the prism is refracted through an angle that depends on the refractive index and the incident angle as described by Snell's law (Eq. 8.8). If the light is composed of several wavelengths, then it will be split up into separate beams because the index of refraction of glass depends on wavelength. These beams can then be separately studied and measured. If light of a single wavelength is incident, it should emerge as a single beam; but if it is incident at a variety of angles, it will be refracted differently. The spectrometer therefore has an entrance slit (perpendicular to the page) and a collimating lens to make sure the beam incident on the prism is parallel. The refracted beam is parallel but quite wide, and if the incident light contained two closely spaced wavelengths, the refracted beams would nearly overlap. The beam is therefore focused by a telescope into images of the entrance slit that are quite small and readily distinguishable from one another. A photographic film placed in the plane of these images records them at intensities corresponding to the intensities at each wavelength in the original beam, as shown in Figure 9.1(b). The various images of the entrance slit are called *spectral lines* because of their appearance on the film. A plot of intensity versus position on the film then gives a spectrum similar to that shown in Figure 7.6.

In many spectrometers the prism is replaced by a diffraction grating and the lenses are replaced by curved mirrors as shown in Figure 9.2. This arrangement has the added advantage that the light is not transmitted by any material and it can therefore be used over a broad range of wavelengths.

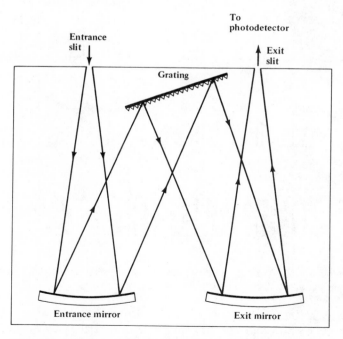

Figure 9.2 Ray-optical diagram of folded grating spectrometer.

Such a device is more compact than the prism spectrometer and allows higher resolution between closely spaced spectral lines because the length of the light path is longer. Furthermore, the film plane is replaced by an exit slit that allows the light of only a single line to pass through it, and the light is then incident on an electronic photodetector (see Appendix G). The grating can be rotated by a slow motor and the spectrum is thereby scanned across the exit slit and electronically recorded. The wiggles and bumps in the recording are also called spectral lines.

The resolution of spectrometers is limited by a number of factors that must be considered together in their design. Clearly the focusing capabilities of the mirrors or the lenses are important to separating closely spaced spectral lines. It is also necessary to have very narrow, straight slits (but if they're too narrow, not enough light will get through, and that which does will be diffracted into a large angle). The grating must have a large number of closely spaced grooves with a minimum of faults, and it should be properly blazed for best efficiency.

The wavelength limits on spectrometers are generally set by the reflection efficiency of the mirrors and grating in the ultraviolet and by spacing between the grating grooves in the infrared. This spacing should be approximately the same as the wavelength and, for wavelengths as long as several microns, a reasonable size grating (10 cm wide) can only have about 10,000 grooves. This imposes an undesirable limit on the resolution of the instrument in the infrared.

B.   Fourier transform spectroscopy was invented to get around some of these difficulties at long wavelengths. The principle is very different from that of traditional optical spectroscopy described above but doesn't involve particularly complicated ideas. Consider the Michelson interferometer shown in Figure 9.3. Light entering from the left is split by a semitransparent mirror into two mutually coherent beams that are reflected from each of the full mirrors. Part of this reflected light from each mirror is recombined by the beam splitter into an exit beam (and part is lost by going back to the source). If the paths taken by the separated beams are the same or differ by an integral number of wavelengths, the exit beam will be strengthened by constructive interference; if the paths differ by an odd number of half wavelengths, the exit beam will be weakened by destructive interference.

Now consider what happens if we send monochromatic light into this interferometer and translate one of the mirrors along the direction of the light beam reflected from it. At various positions the destructive or constructive condition will be met, and at others the situation will be partway between. The strength of the output beam will vary sinusoidally if the mirror is moved at constant velocity. If we compute the Fourier transform of this

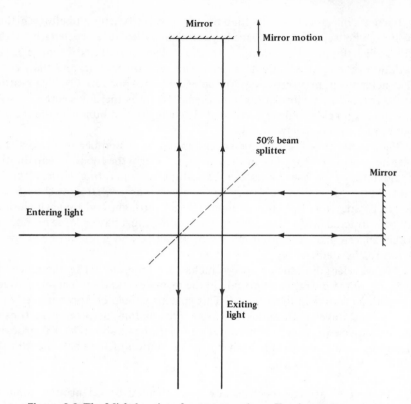

Figure 9.3 The Michelson interferometer used as a Fourier spectrometer.

sinusoidal variation (see Part 7.2), we find a single spike that represents, of course, the spectrum of monochromatic light. If we change the wavelength of the incident light, the modulation of the exit beam by the moving mirror will still be a sine wave whose Fourier transform is a single spike, but it will occur at a different frequency because the wavelength is different. If we superpose two or more beams of light and send them into the interferometer, the output intensity may vary with mirror position in some complicated way (as shown in Figure 7.5 and 7.6) but the Fourier transform will still display the spectrum of the incident light.

Fourier transform spectroscopy is named because it is necessary to transform the output of the spectrometer (in this case interferometer) in order to see the spectrum. It is usable at any wavelength, but the suitability of conventional optics and the difficulties of precision movements generally favor traditional spectroscopy for the visible and shorter wavelengths. It is extremely important however, for longer wavelength studies, and as we shall see, most molecules of biological interest absorb and emit infrared light.

Part 2 : Molecular Structure

Most of what we know about the structure of atoms and molecules has been learned from the study of the light they emit and absorb (optical spectroscopy). The electromagnetic radiation (visible light, infrared, ultraviolet, radio frequency, etc.) derives classically from the orbital acceleration of charged particles within the molecules. Since the motion of the nuclei and electrons is determined by their number and kind in the molecule, each molecule has its own particular set of frequencies of electromagnetic radiation. Once this set of frequencies has been identified, it can be used as a probe for the presence of a particular molecule as well as for any perturbations on it.

A. The atoms that group together to form molecules are bound to each other by forces between the nuclei and electrons. A molecule can be stable against decomposition if these binding forces can pull the atoms back together after a collision or other kind of disturbance causes them to move out of their normal equilibrium orbits. The atoms are therefore elastically bound in a molecule and can vibrate about their equilibrium positions.

It is customary to discuss the potential derived from the force between the atoms of a molecule. This potential $U(r)$ is given by

$$F = \frac{-dU(r)}{dr} \tag{9.1}$$

and is generally plotted for discussion by choosing the interatomic potential to be zero when the atoms are so very far apart that they no longer interact to any significant degree. In Figure 9.4 we show that this potential is zero for very large r and decreases as r becomes smaller. The resulting positive slope implies that the force between the atoms is negative and therefore attractive. If the atoms come too close together, they "bump into each other" and there is a strong force pushing them apart. Hence the slope of the curve is negative (positive force) for small r.

For intermediate values of r there is a dip (potential well) in the curve. At the bottom of the well the slope is zero; hence the force is zero, and this represents the equilibrium value of the interatomic separation. If the interatomic distance is slightly displaced to either side of the well, there will be a restoring force and the molecule will vibrate just as a frictionless bead might oscillate in the potential well if it were under the influence of a gravitational force. If the atoms are forced so close together that they rise on the potential curve to a positive energy, then enough work has been done on the molecule so that it will fly apart when the atoms are released. The work required to do that is just the depth of the well and is called the *binding energy* of the molecule.

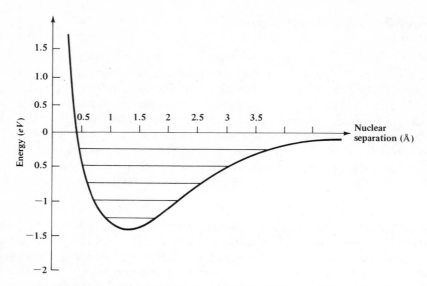

Figure 9.4 Plot of the electrostatic energy of two atoms as a function of their separation. The existence of a minimum in the curve shows that they will bind together to form a molecule. There are several bound vibrational levels indicated by the horizontal lines.

As part of the introduction to sound waves in Part 7.1 we studied resonance and normal modes of vibration. Our study of molecular vibrations continues along the lines established there and is concerned with a description in terms of normal modes. A diatomic molecule can have only the vibrational mode that is along the internuclear axis. The frequency of the vibrations is determined by the masses of the nuclei and strength of the force pushing the nuclei back toward their equilibrium position (the spring constant k of Chapter 7). Since this force comes from interactions between the nuclei and electrons, we can use the measured frequencies (and known masses) to learn about the electronic structure of molecules. The amplitude of the vibration is determined by whatever mechanism starts the vibration and does not usually provide much information about molecular structure.

If the vibration is driven by an oscillatory force such as the time-varying electric field of incident electromagnetic radiation, the molecule will exhibit resonant behavior just as the mass-spring oscillator of Chapter 7. The damping or friction force is usually provided by energy loss through radiation, although the amplitude of oscillation can be limited because too large an amplitude will cause the molecule to fly apart. The molecule will vibrate at some amplitude determined by its binding forces and the parameters of the driving field, and the energy of the vibration will be a monotonic function of the vibration amplitude.

During the 1920's it was learned that the detailed internal behavior of atoms and molecules cannot be completely described by classical Newtonian mechanics. In addition to the classical concepts, it was found necessary to add certain other constraints on the variables defining the motion of particles. The quantitative description of the motion is called *quantum mechanics*, and in the case of vibrating molecules it is manifest in a quantization of the vibration energy. The frequency f of the vibration in the quantum mechanical picture is unchanged from the classical frequency, but the amplitude can assume only those particular values that correspond to energies E_v such that

$$E_v = (v + \tfrac{1}{2})hf \qquad (9.2)$$

where v is an integer and h is approximately 6.6×10^{-34} Joule-sec (Planck's constant). The molecule is therefore NOT allowed to have any arbitrary vibration energy but only those energies specified by Eq. 9.2 with various values of the integer v.

We represent these energies in Figure 9.4 with horizontal lines spaced by the energy difference hf. It is clear that there are only a finite number of allowed energy states and that they are each separated by the same amount. Solutions to the *Schroedinger equation*, which describes the quantum mechanical behavior of atoms and molecules, consist of the allowed normal modes of vibration for this system. They are called *eigenfunctions*, which means natural or normal solution, and each one has an associated *eigenvalue*, which corresponds to the energy of the molecular state of motion, E_v.

Of course the molecule's vibrational energy can change, but when it does, it can only be by an amount that is an integral multiple of the quantity hf. If the change in v is one, we say it has changed by one *quantum*; if it is two, we say two quanta; etc. Furthermore, the energy absorbed from or emitted into the electromagnetic radiation field when these changes take place can only be in discrete amounts of energy (also called quanta), thus giving rise to an emission or absorption spectrum that consists of discrete frequencies corresponding to the energies of the quanta. Since f is different for different molecules (masses are different), each molecule has its own characteristic vibrational spectrum consisting of a series of frequencies (called spectral lines) at particular frequencies. For most molecules, these frequencies are about 10^{13} Hz and the spectrum falls in the infrared.

The discussion of vibrational spectra of diatomic molecules has to be modified and extended somewhat for polyatomic molecules. If the three atoms of a triatomic molecule form a straight line, there are four normal modes of vibration. Two of them are bending modes (in perpendicular directions) in which the distance between the end and center atoms doesn't change, and the other two are oscillatory in which the three atoms move along a straight line. As the number of atoms in a molecule increases, the number of normal modes of vibration increases rapidly. For example, formaldehyde

(H_2CO, four atoms) has six different normal modes of vibration. The vibrational spectra of polyatomic molecules are extremely complicated because they can make transitions in which energy is transferred between the normal modes. For example, if a molecule has three normal modes, its energy is $E(v_1, v_2, v_3)$ and there is a nice spectrum as only one of the v's changes, but it's quite likely that more than one of the v's will change resulting in an energy change that depends on the frequencies of more than one of the normal modes of oscillation. Such transitions will produce spectral lines interspersed with those of the simple spectrum and indistinguishable from them. The result is a large number of closely spaced spectral lines whose frequencies are related in a complicated way. It is a difficult job for the spectroscopist to unravel the spectrum of even the simplest polyatomic molecules.

B. In the foregoing discussion of the energy levels of molecules we have considered only the vibrational motion. Molecules can rotate, of course, and associated with this rotation is a certain amount of kinetic energy. This energy comes from the component of the motion of the atoms associated with rotation rather than vibration and is given by

$$E = \tfrac{1}{2}I\omega^2 \tag{9.3}$$

where I is the moment of inertia of the molecule about the axis of rotation and ω is the angular frequency in radians per second.

The rules of quantum mechanics also require that angular momentum L is quantized and for a rotating molecule we write

$$L = I\omega = \frac{nh}{2\pi} \tag{9.4}$$

where n is any integer and h is Planck's constant as before. It is easy to see the effects of angular momentum quantization on rotational energy. We combine Eqs. 9.3 and 9.4 and find

$$E_n = \frac{1}{2}I\omega^2 = \frac{n^2h^2}{8I\pi^2} \tag{9.5}$$

for the allowed rotational energies. For a typical molecule, oxygen for example, this becomes

$$E_n = 6 \times 10^{10}n^2h \tag{9.6}$$

which means the frequency of electromagnetic radiation of this energy is 6×10^{10} Hz or in the microwave region.

The energy associated with molecular rotation is typically 1000 times smaller than the vibration energy and therefore difficult to show on a graph like Figure 9.4. Nevertheless it is customary to exaggerate the separation of the rotational energy levels for demonstrative purposes and display them as shown in Figure 9.5. It is clear that a molecular transition between two

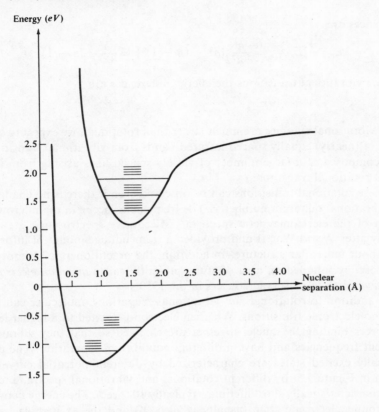

Figure 9.5 Plot of energy versus separation as in Figure 9.4. A curve for an electronically excited state is shown, as are several rotational levels in each of the vibrational levels.

vibrational levels can occur at a large number of closely spaced frequencies corresponding to changes in the rotational state of the molecule. These frequencies are so closely spaced that the limited resolution of early spectroscopy failed to resolve them into individual spectral lines [see Figure 9.1(b)] and the vibrational spectra were thought to be continuous bands. The name *band spectra* is retained today even though we can readily resolve many rotational transitions.

We can write the energy of a molecular state as

$$E_{n,v} = \frac{h^2}{8\pi^2 I}n^2 + hf\left(v + \frac{1}{2}\right) \tag{9.7}$$

and calculate the energies associated with various transitions between states. We then use $E = h\nu$ and $\nu\lambda = c$ to find the wavelengths and the expected spectrum. We find that, for rotational transitions with $\Delta n = 1$, the energy differences are

$$E_{n,v} - E_{n-1,v} = \frac{h^2}{8\pi^2 I}[n^2 - (n-1)^2] = \frac{h^2}{8\pi^2 I}(2n-1) \qquad (9.8)$$

and for vibrational transitions the energy differences are

$$E_{n,v} - E_{n,v-1} = hf\Delta v = hf. \qquad (9.9)$$

Since vibrational energies are much larger than rotational, we expect to see a series of (nearly) equally spaced infrared bands from vibrational transitions, each composed of a large number of closely spaced lines arising from individual rotational transitions.

These rotational transitions can be observed even if there is no change in the vibrational quantum number v. The frequencies occur in the microwave region of the electromagnetic spectrum. Microwave spectroscopy grew up shortly after World War II and provided a tremendous amount of information about molecular structure. In addition, the practitioners of microwave spectroscopy founded an area of study presently called *quantum electronics* that gave birth to masers and lasers in the 1950's.

In addition to rotational and vibrational excitations, molecules can also undergo electronic transitions. When an electron is excited to a higher level, the forces binding the nuclei together are changed so that they vibrate at different frequencies and have a different equilibrium separation. The electronically excited states are characterized by different potential curves as shown in Figure 9.5, by different rotational and vibrational spectra, and by their characteristically short lifetimes (typically 10^{-8} sec). The energy between the ground electronic state (usually labeled X) and the excited electronic states (usually labeled A, B, C, etc.) is typically several electron volts and corresponds to ultraviolet wavelengths of light. In order to calculate the spectrum, simply add the electronic excitation energy, E_A, E_B, etc., to Eq. 9.7 and proceed as before. Light emitted by molecules undergoing electronic transitions has a spectrum that generally shows several ultraviolet vibrational bands composed of a large number of unresolved rotational transitions. Under high resolution, these bands also show individual rotational lines with energy separations given by Eq. 9.8. It is clear from following the potential curves of electronically excited states to large values of r that the separated atoms have an energy that is not zero but at least one of them is in an excited state.

High resolution studies of individual bands show very many spectral lines that have to be sorted out in order to determine which rotational transitions produced them. The wavelength intervals are used to determine the molecular moment of inertia and hence the separation between the nuclei. This data provides further information about the structure of the molecule.

If one of the atoms in a molecule is replaced by its isotope, the chemical properties of the molecule are changed very little because the number of electrons is unchanged and their orbits are only slightly perturbed. On the other hand, the different mass of the isotope changes both the vibrational *and* rotational spectrum of the molecule in rather substantial ways. It is possible to use a tunable laser to excite levels of a molecule containing one isotope but not another and to react selectively, dissociate, or ionize these excited molecules. One can then collect the products that now contain only a selected isotope. This procedure forms the basis of laser isotope separation that is a very important way of providing isotopes for tracer studies.

Modern, narrow-band, tunable lasers have advanced the techniques of optical spectroscopy so much that laser spectroscopy is now used routinely for high resolution and high precision molecular measurements. The result is a major expansion in the study of molecular structure which includes molecules of biological importance. Various kinds of non-linear methods, including saturation, Lamp-dip, and two-photon spectroscopy have enabled researchers to study optical transitions without the loss of resolution that results from Doppler broadening. Polarization spectroscopy has facilitated identification of levels. Time resolved spectroscopy using pulsed lasers has provided accurate measurement of molecular parameters. Very short light pulses (picoseconds) are revealing details of certain reactions such as photosynthesis. Lasers will surely produce a tremendous amount of new information about biologically interesting molecules in the next few years.

We have discussed the basic structural features of simple molecules and the methods used to determine them. The extension to more complicated molecules is tedious but straightforward. We shall see that the confluence of optical spectroscopy and other spectroscopic techniques can provide information about the structure of simple amino acids and other polyatomic molecules of biological importance.

Part 3: Resonance Spectroscopy

A. Along with its charge and mass, an electron also has magnetic properties associated with its spin angular momentum. In particular, the electron has a magnetic moment $\mu \cong 2 \times$ (Bohr magneton) that can be oriented in a magnetic field in such a way that there are two separate, distinct energy states. These are commonly called spin up and spin down, and the electron can make a transition between these energy levels that is called a *spin flip*. The energy associated with the transition is directly proportional to the magnetic field at the electron, which is the sum of an applied, external magnetic field and local magnetic field produced by the nuclei and other electrons in the molecule. Study of these magnetic energies is called *spin resonance* and is usually done in applied fields between 3000 and 4000 gauss where the primary contribution to the spin flip frequency is from the applied field and the frequency is in the 10 GHz region. The molecular structure causes small deviations from the applied field to appear at the electrons so that their spin flip frequencies are slightly different. The frequencies are calculated from the energies and are given by

$$\text{frequency} = \frac{E}{h} = \frac{\mu H}{h} = \frac{\mu}{h}(H_{\text{applied}} + H_{\text{local}}) \tag{9.10}$$

where H is the magnetic field. If different electrons see different local magnetic fields, there will be several frequencies where the resonance condition is met and the transitions will occur. By measuring the frequencies it is possible to study the magnetic environment of various electrons in a molecule and, therefore, learn more about the molecular structure. The technique is called *electron spin resonance* (ESR) or *electron paramagnetic resonance* (EPR).

When electrons are placed in a magnetic field with no prior sample preparation, it is equally likely that any given spin will orient itself up or down, and we might therefore expect an equal number of electrons in either orientation in a macroscopic sample. If the sample is subjected to electromagnetic radiation at the right frequency for ESR, some of the electrons will be switched from up to down and vice versa, but there will be no net change in either the sample or the radiation field. On the other hand, if there are more electrons in the lower energy state, there will be a net absorption of electomagnetic energy from the radiation field as it equalizes the electron population between the spin up and spin down states. Furthermore, the amount of energy absorbed will increase as the electron population difference increases, so that large differences produce strong absorptions.

The population differences are produced by thermal effects. At very high temperatures, we expect so much thermal agitation from other molecules in the sample and from the blackbody radiation field that the two spin levels are equally populated. At extremely low temperatures we expect all the electrons to be in their lowest energy state because they would all decay down to that state and stay there. At all temperatures the ratio of the populations is found from thermodynamic considerations to be

$$\frac{N_{\text{upper}}}{N_{\text{lower}}} = e^{-E/kT} \tag{9.11}$$

where k is the Boltzmann constant, T is the absolute temperature in °K, and E is the energy between the upper and lower levels under study. This expression has the right behavior at temperature extremes and gives the relative population at laboratory temperatures. It is clear that there are certain advantages to doing ESR at very low temperatures.

The electromagnetic radiation equalizes the population of the upper and lower levels leaving no population difference to absorb more radiation. Therefore it is possible to observe an ESR signal only if the thermal processes that relax the population toward thermal equilibrium (Eq. 9.11) are at work to maintain some population difference. It is necessary for the thermal relaxation process to occur in a time that is shorter or comparable to the time it takes an electron to make a transition. If this condition is not met, the result is saturation of the ESR signals, which will make them very difficult to see.

The signals are observed by monitoring the absorption of the 10-GHz radiation. This microwave energy (wavelength about 3 cm) is typically transmitted via waveguide to a sample placed in a resonant cavity whose size is dictated by the wavelength. The cavity is between the pole pieces of a large electromagnet. Usually the applied magnetic field is continuously varied (swept) through the region where the transitions are expected and the amount of power reflected back from or transmitted through the cavity is monitored. When the ESR resonance condition is met, microwave energy is absorbed by the sample and the monitoring apparatus records the change. The results of an ESR experiment show a series of peaks and is called an ESR spectrum in analogy with optical spectra. The individual signals are even called *spectral lines*. A characteristic ESR spectrum is shown in Figure 9.6.

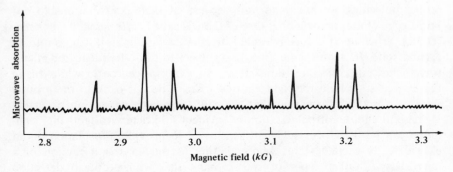

Figure 9.6 An ESR spectrum. The peaks at different applied fields indicate that there are electrons in different local fields.

B. Many atomic nuclei have magnetic moments arising from their spin angular momentum in much the same way as the electron magnetic moments, but the nuclear moments are generally several thousand times smaller. Nevertheless, when nuclei are subject to an applied magnetic field, the finite temperature of the sample produces a population difference between upper and lower states, and it is possible to observe spin flip transitions of the nuclei. This experimental technique is called *nuclear magnetic resonance* (NMR) and can be used for careful study of the magnetic environment of nuclei. Of course the small magnetic moments result in smaller energy splittings so that the population differences are much smaller and the resonance frequencies are much lower (for the same magnetic field).

In spite of the apparent difficulties of NMR relative to ESR, NMR has contributed a vast amount to our knowledge of molecular and crystal structure especially in biological materials. Because NMR resonances occur in the radio-frequency range, very sensitive detection circuits can be made to observe the nuclear absorption of energy. In addition, there are other electronic techniques involving radio-frequency transmitters and receivers that can enhance the sensitivity of NMR substantially. Ths result is that commercial manufacturers sell entire NMR systems that include the electromagnet, power supply, radio-frequency equipment, detection electronics, etc., in a single package appropriately called an NMR spectrometer (prices start at about $30,000). On the other end of the scale, it is possible to make an NMR apparatus for undergraduate laboratories that costs about $50.

Nuclei also have quadrupole moments that behave in many ways like their magnetic (dipole) moments. It is possible to study *nuclear quadrupole resonance* (NQR) or *paramagnetic quadrupole resonance* (PQR).

C. There are a number of other types of resonance spectroscopy that have yet to make substantial contributions to our knowledge of biologically interesting molecules but are important because of their contributions to our knowledge of atomic structure. One of these is called *optical double resonance* (ODR). When an atom or molecule is irradiated with light, it can go into an excited state. If, at the same time, it is subject to radio-frequency or microwave fields, magnetic transitions similar to spin flips can occur while it is in the excited state, and the light emitted when it decays back to the ground state carries the information that the radio-frequency transition occurred. By careful choices of polarization of incident and detected light, this technique can be used with sufficient sensitivity to provide information about electronically excited states whose lifetime is only a small fraction of a microsecond. Other types of spectroscopy that utilize optical detection include, but are not limited to, level crossing and optical pumping.

Part 4: Microscopy

One of the purposes of a microscope is to enable visual examination of cells and other small structures of biological interest. There are many different kinds of microscopes with different limitations and uses, but the most common by far is the optical microscope and we shall begin with it.

A. Our ability to see small details with the unaided eye is limited to objects that subtend an angle larger than about 2×10^{-4} radian as discussed in Part 8.5. It is clear from there that visual acuity is limited by the angular size of the object and not by its actual size. For example, you can read this page when it is 30 cm away but not when it is 30 m away. This is not because the size of the individual letters has changed, but because they subtend a smaller angle. Since the ability of our eye to focus is limited to objects at least 20 cm away (see Part 8.3), the smallest detail we can expect to resolve with the naked eye is about 40 microns (an average human hair is about 80 microns diameter). Most cells are about 1 or 2 microns in diameter and therefore invisible to the naked eye. The function of an optical microscope is to allow us to see objects when they are closer to the eye than 20 cm and to provide some angular magnification as well.

The simplest optical microscope is a single lens magnifier that is hand held in front of the eye. It adds to the focusing power of the eye in accordance with Eq. 8.13 enabling it to focus on an object that is closer than 20 cm. We combine Eq. 8.13 and 8.11 to find that a lens of focal length f brings the closest focus from a (about 20 cm) to

$$a' = \frac{af}{a + f}. \tag{9.12}$$

Since f is usually considerably smaller than a, a' is approximately equal to f and the apparent angular size of an object is increased by $\sim 20/f$, the magnification of a simple magnifier.

At first thought it appears that considerably more magnification might be achieved if the magnifier were held at approximately its focal length away from the object resulting in a very much enlarged image. We examine this idea beginning with Eq. 8.11, which tells us that the location of the image is at

$$s' = \frac{sf}{s - f} \tag{9.13}$$

and is magnified by an amount

$$m = \frac{-s'}{s} = \frac{-f}{s - f}. \tag{9.14}$$

We can choose s to be just less than f resulting in a very large virtual image located at a great distance. The image size h' would then be m times the object size h, and the angular size of the image would be

$$\frac{h'}{s'} = \frac{mh}{s'} = \frac{h}{s}. \tag{9.15}$$

The largest angular size that can be achieved with the naked eye is about $h/20$ since the eye can not focus on an object choser than about 20 cm. The angular size of an image from the single lens in this case is h/s and the magnification is therefore $20/s \cong 20/f$ (see Eq. 8.15) since we have chosen $s \cong f$. The magnification is not any larger. The magnification doesn't depend on the distance between the lens and the eye as long as the distance between the lens and the object is slightly less than f, but the field of view is much larger if the lens is held closer to the eye. It therefore makes sense to hold the lens comfortably close to the eye and to move the object into proper focus.

On the other hand, we can choose s to be slightly larger than f. In this case there is a large real image formed between the eye and the lens. This image must be at least 20 cm from the eye or we can't focus on it. Furthermore, the distance from the eye to the object should not exceed an arm's length or we would need something to hold it. We let arm's length be L and the distance from object to image be

$$s + s' = L - 20 = x. \tag{9.16}$$

We find that

$$\frac{1}{s} + \frac{1}{s'} = \frac{1}{s} + \frac{1}{x - s} = \frac{1}{f} \tag{9.17}$$

which can be solved for s, and we substitute into Eq. 9.14 to find

$$m = \frac{s'}{s} = \frac{x - s}{s} = \frac{1 \mp \sqrt{1 - z}}{1 \pm \sqrt{1 - z}} \tag{9.18}$$

where

$$z = \frac{4f}{x}. \tag{9.19}$$

In order for the magnification m in Eq. 9.18 to be real, it is necessary that

$$z = \frac{4f}{x} < 1. \tag{9.20}$$

It is clear that if z is only slightly less than 1, the magnification will be only slightly different from unity whether we take the upper or lower signs in Eq. 9.18 because the radical is nearly zero. Conversely, if z is very small compared with unity, the upper signs lead to a vanishing small magnification, while the lower signs lead to a magnification of x/f. When we consider that L is typically about 50 cm, which means x is not larger than 30 cm, the practical limit to magnification in this case is $30/f$, not much larger than the value $20/f$ for the case when the lens is close to the eye. Working at full arm's length results in awkward positions, poor stability and focus, and a very much smaller field of view in exchange for only 50% increase in magnification. That is why it is not usually done.

B.   We have expended considerable effort studying the normal operation of the single lens magnifier because it is important. Its magnification is approximately $20/f$ but cannot be increased without limit for practical reasons. When we try to make f very small, it is clear from Eq. 8.10 that the radii of curvature of the surfaces must become small. For $R_a = -R_b = R$ and $n = 1.5$, we find $f = R$. It is clear that we cannot make R, the radius of curvature, any smaller than half the diameter of the lens itself thus limiting the focal length of a lens to be greater than half its diameter. Since lenses for practical use are limited to being larger than about 10-mm in diameter, a practical limit to the magnification of a single hand-held lens is about 40.

In order to achieve magnification greater than this it is necessary to use two or more lenses to form a compound microscope. One of them, the objective, is held very close to the object and produces an enlarged, real image of magnification given by Eq. 9.18. Usually f is very small compared with x (typically $f = 1$ mm) and the image is formed fairly close to the eye. This image is then examined by another lens of focal length about 2 cm placed close to the eye and used as a single lens magnifier. The two lenses are placed at opposite ends of a tube, and the entire apparatus is precisely movable with respect to a fixed, mounted object, as shown in Figure 9.7.

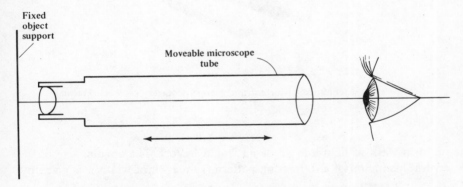

Figure 9.7 Schematic illustration of a compound microscope. The objective and eyepiece are mounted in a tube that is moved relative to the object being viewed.

Compound microscopes can be purchased from a large number of manufacturers with a wide range of price and quality. The example shown in Figure 9.8 has a rotatable turret in which the first lens (objective) can be interchanged for others of different focal lengths.

The magnification of a single lens magnifier is limited by practical considerations, but the magnification of a compound microscope is limited by the fundamental physical property of waves called *diffraction*. The diffraction theory of image magnification by a microscope was developed by Abbé and is presented in many texts (poorly in some). The essential physical idea is that there can be no image formed if the diffraction pattern of the object spreads out until it fills the entire aperture of the lens, because there would then be a

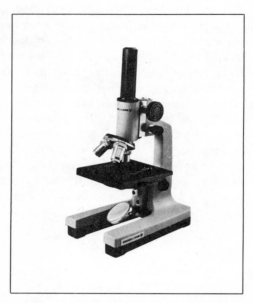

Figure 9.8 A commercial turrent microscope. Courtesy of W.H. Freeman and Sons, San Francisco.

uniform wash of light and all detail would be lost. The angular size of the central maximum of a diffraction pattern from a slit of width w is given by

$$\sin \theta \cong \frac{\lambda}{2w} \tag{9.21}$$

where λ is the wavelength. Since $\sin \theta$ can never exceed unity, it is clear that any w smaller than $\lambda/2$ will produce a diffraction pattern that cannot be collected by any microscope objective and will therefore be unresolvable. In practice, no microscope objective can accept a cone of light that has $\sin \theta = 1$, but there is little problem at $\sin \theta = 0.8$.

There are several tricks that can be used to resolve detail at a slightly smaller scale. One is to immerse the object in some fluid medium whose index of refraction is considerably larger than unity so that the effective wavelength is reduced. There are special oils chosen for this purpose, and the specially designed objective is described as an oil immersion objective.

Another trick is to select only the short wavelength components of visible light. Microscope illuminators are usually blue or purple for this reason.

There are a number of objects that cannot be easily seen in an ordinary light microscope because they are neither opaque nor do they refract light strongly. This problem is sometimes overcome by staining the sample, but specially designed microscopes have been built to make it easier to study them without staining. One such scheme is to put the object between crossed polarizers. In this "polarization microscope" the visibility of an object is determined by the ability of its substance to rotate the plane of polarization (birefringence) in order to permit light to pass the second polarizer. The objects appear brightly lit against a dark field.

Another method illuminates the objects obliquely so that none of the original illuminant enters the objective. An optical system produces a hollow cone of light, some of which can be scattered by the object into the objective. In this "dark field microscope," the object appears brightly illuminated against a dark field just as in the polarization microscope.

An interference (or phase contrast) microscope utilizes the phase shift as light passes through an object to produce an interference pattern that is magnified by the microscope. In this case the object appears as a change in the array of fringes that comprises the interference pattern.

Increasing the visibility by using polarization, dark field, and interference, does not increase the resolution of detail. These techniques may, however, be combined with each other and with resolution enhancement such as oil immersion and shortwave illumination, to produce other specialized microscopes. There are many dozens of kinds to choose from, and the particular application must be well understood before selecting one of these instruments.

C. All optical microscopes are subject to the fundamental resolution limits described above. The only way to achieve substantially greater magnification is to use radiation of substantially shorter wavelengths. Ultraviolet microscopes have been built, but the image must be recorded photographically. X-Ray microscopes have been built, but X-Ray lenses have such poor optical quality (they are basically geometrical devices) that there has been no significant improvement from them either. Ultrasonic microscopes have not yet achieved short enough wavelengths to be of value, but they do offer the special capability of seeing below the surface.

The real breakthrough in microscopy comes from the electron microscope. Electrons can be focused by specially designed metal structures with various voltages on them as well as by magnetic fields from current-carrying coils. Modern electron microscopes can resolve details as small as 3 or 4 Å, 1000 times smaller than the best optical microscopes, and they have not yet reached a fundamental limit such as diffraction. They are presently limited by the small acceptance angle of about 0.01 radian of available electron lenses. For 50-keV electrons, the wavelength is less than 0.1 Å but the acceptance angle limits the resolution to $w = \lambda/2 \sin \theta \cong 5$ Å in accordance with Eq. 9.21.

The scanning electron microscope (SEM) offers a radical alternative to conventional microscopy. It is clear from Eq. 9.13 (which applies to electron optics as well as conventional optics) that an image formed at a distance s' will be in focus only if there is very good control over the lens-to-object distance s. If the object has any thickness, only part of it will be in focus and the rest will produce a blur. This limited depth of field, common to all microscopes, requires that the sample be prepared as an extremely thin slice. The SEM has a much greater depth of field that allows completely different views of an object.

In Figure 9.9 it is shown that a SEM does not really form an image of the object. Instead, a very well-focused beam of electrons is scanned across the surface and ejects secondary electrons as it is swept. These secondary electrons are collected and amplified into a current that controls the brightness of a second electron beam writing on a conventional TV screen. The beam writing on the TV screen is scanned synchronously with the SEM beam and

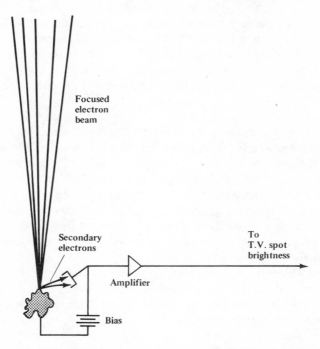

Figure 9.9 Schematic illustration of the workings of a scanning electron microscope. It is not a microscope in the conventional sense.

produces an image that shows the secondary emission capability of the sample. Since this varies with the contours and angles of the object, an image is formed. The SEM has tremendous depth of field because the angular sensitivity is determined by the sweeping beam before it reaches the object. Its magnification can be controlled electronically by varying the sweep, and its image can be transmitted on cables as a conventional TV image. A sample SEM picture is shown in Figure 9.10. The potential of this relatively new microscope is still being explored both in research and as a teaching device.

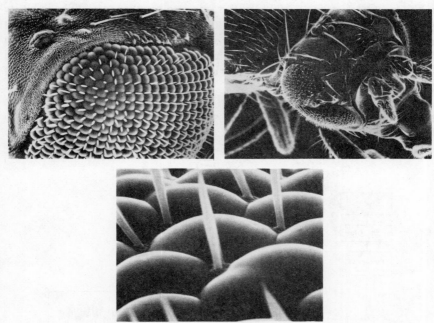

Figure 9.10 Photographs of the screen of a scanning electron microscope. Courtesy of T. Hayes and H. Hartman, Lawrence Berkeley Laboratory, University of California, Berkeley.

Part 5: Electronic Signals and Noise

Almost all instruments used in modern scientific research depend on electronics in some way. These instruments span the range of sophistication from a stabilized light source for taking photomicrographs to a complex computer system for running experiments and taking data. Furthermore, all electronic signals are characterized by some level of unwanted fluctuations in voltage or current that are not under the control of the experimentor and are not related to the experiment in any direct way. These fluctuations, called *noise* by analogy to their acoustical counterpart, are often very much smaller than the signal of interest and are generally ignored. We say the *signal-to-noise ratio*, *S/N*, is large. On the other hand, there are many cases where the signal is so small that the noise is always dominant, and it requires special techniques to enhance the *S/N* so that the signal can be detected in the presence of the noise. Electronic instruments for doing this job are the topic of this section.

A. The simplest electronic device for enhancing *S/N* is a resistor-capacitor (*RC*) network used to smooth the signal. In many cases the signal of interest is the variation of one variable with respect to another (e.g., light intensity versus wavelength). If the speed at which the independent variable is swept under experimental control is very slow, the noise fluctuations may occur in a time much shorter than that required to sweep over the range of interest, and the smoothing of an *RC* network can result in considerable enhancement of the *S/N*.

In Figure 9.11 a signal derived from some detector is displayed both before and after *RC* filtering. The capacitor acts as a low resistance path for high frequencies in accordance with

$$Z_c = \frac{1}{2\pi f C} \tag{9.22}$$

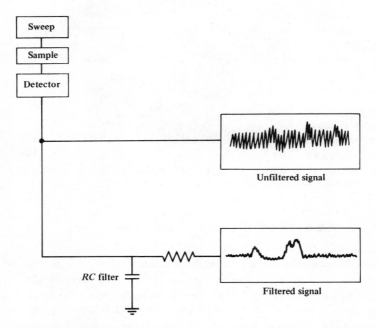

Figure 9.11 Block diagram of an *RC* time constant filter for signal-to-noise enhancement.

where Z_c is the capacitive reactance in ohms, C is the capacitance in farads, and f is the frequency in hertz. The resistance R in Figure 9.11 does not vary with frequency so that high frequencies are preferentially shunted to ground by the capacitor and low frequencies are preferentially delivered to the data acquisition system (in this case a display) by the resistor. If the independent variable is swept slowly enough, and the product RC is large enough, a substantial increase in S/N can be obtained.

The essential characteristics of S/N enhancement by RC filtering are determined by the choice of sweep speed and time constant (RC). If the sweep speed is sufficiently fast that a feature of interest is traversed in a time shorter than RC, it will be filtered out along with the noise and therefore not be observed. If the sweep speed is too slow, the experiment will take too long to be practical. It is necessary to study the nature of the noise that is mixed with the signal and to choose appropriate averaging times. All the S/N enhancement devices we shall study depend on some kind of time averaging or filtering of the noise and, therefore, require careful consideration of the noise spectrum, the time constant, and the sweep speed.

B. One class of instruments, called *signal averagers*, extracts signals from noise by time averaging, just as the simple RC filter does, but allows for multiple sweeps, digital storage, and easy communication with computers. In fact, many computers are used as signal averagers and many commercial signal averagers are simply dedicated computers. The development and widespread use of low cost microprocessors will contribute to erasing the distinctions between signal averagers and computers.

All signal averagers depend on the assumption that the noise obscuring the desired signal is not correlated with it and is therefore random. A signal averager is designed to record an electronic signal delivered to it during a single sweep of some indpendent variable and store the signal in its multi-channel memory. The independent variable is then swept again, and the incoming signal is added to the memory. Repeated sweeps allow many "copies" of the noisy signal to be added together resulting in an enhancement of any real signal and a relative reduction of the noise by averaging.

The major functional parts of a signal averager are the sweep control and synchronization electronics, signal processing and averaging circuits, memory and memory control hardware, and readout and display systems. Since each of these basic components is rather complex, commercial signal averagers are expensive (as high as $10,000) and sophisticated devices. Nevertheless they form the heart of the research apparatus in many laboratories and are becoming more and more common. For this reason it pays to discuss their capabilities further.

In Appendix G we studied the random walk problem and it was shown that adding together the result of a large number of independent random processes produces a parameter that increases as the square root of the number of steps. That is, if we add together n separate steps of random direction but equal magnitude, we expect that the net displacement should be not zero but approximately $\sqrt{n}$ times the step length. Other arguments based on the random (uncorrelated) nature of noise lead to the idea that if n signals with noise N and signal S are added together, the total noise will be $N\sqrt{n}$ and the signal will be nS resulting in a final signal-to-noise value that is $\sqrt{n}$ times larger than the original value.

There is one particular case where this $\sqrt{n}$ measure of noise can be shown rigorously. If a stream of particles (products of a radioactive decay, photoelectrons from a laser detector, etc.) arrive at a rate whose long-time average is S per second, then the number of these expected in any time interval T is simply ST. If repeated measurements are made during several time intervals, the results will fluctuate of course, but the standard deviation of the several measurements whose average is ST will be $\sqrt{ST}$; that is, the noise associated with a certain number of counts is the square root of that number, and the signal-to-noise value is then

$$\frac{S}{N} = \frac{ST}{\sqrt{ST}} = \sqrt{ST}. \tag{9.23}$$

This is rigorously correct for certain special cases where our ignorance of the exact time of emission of the detected particles allows us to make certain statistical statements about their time distribution (called *Poisson statistics* or Poisson distribution). Since many biological experiments rely on detection of radioactive decay or counting of photoelectrons from optical detectors, the $\sqrt{n}$ idea permeates the interpretation of the data.

The $\sqrt{n}$ measure of statistical fluctuation is often generalized to areas where its use is unjustified and may even be wrong. Most scientists will be inclined to reject observations if their magnitudes are smaller than some measure of the square root of the sample size and to accept them if they are larger, but it is easy to make errors this way if one is not careful. For example, changing the gain of an amplifier or multiplying all the data in a table by some constant will surely change $\sqrt{n}$ but will not change S/N in any way.

With the warnings above in mind, it is possible to predict the noise, S/N, and enchancement by signal averager in a given experiment. For example, if one wants to study the influence of a particular drug in the uptake of iodine by the thyroid using a radioactive tracer, one simply calculates how much of a change can be detected in a given time. If the average rate at which one detects decay gammas from I^{131} is S per/second, then the fluctuations in any time interval T is $\sqrt{ST}$, which determines the precision of a measurement. For example, a 1% change in count rate could be reliably detected in $10^4/S$ seconds, or a few times longer than that. If the drug's influence on the uptake rate were to be studied as a function of body temperature, for example, it is clear how one would calculate the length of time needed to study each temperature and therefore the rate at which to sweep the temperature and the signal averager.

C. There is another very common technique of S/N enhancement called *phase-sensitive detection*, which is widely used in magnetic resonance (see Part 9.3). In this method the independent variable is not only swept slowly through the resonance but also modulated over a very narrow range at a frequency usually in the audio. The desired signal is also modulated because it is correlated with the independent variable, but the noise is not affected. The detector output is fed to a phase-sensitive detector (also called a *lock-in amplifier*) which is simply a very narrow band amplifier which passes (and amplifies) waveforms which are precisely at the modulation frequency and in phase with the modulation. Since most of the noise does not meet this criterion, it is not amplified, but the signal is amplified resulting in an increase of the S/N.

Consider the resonance signal shown in Figure 9.12(a). The effect of the modulation creates a modulated signal at the same frequency and of fixed phase. The amplitude of the signal is proportional (approximately) to the slope of the resonance curve, and the phase reverses with the slope as shown. At the peak of the resonance curve, the amplitude of the modulated signal is zero. A lock-in amplifier will produce the output waveform shown in Figure 9.12(b). It is largest where the slope is high and zero at the center: it goes negative where the phase reverses relative to the modulation waveform. A spectrum with several lines will appear as a series of these dispersion-shaped curves as shown in Figure 9.12(c).

A block diagram of a lock-in amplifier is shown in Figure 9.13. The modulation, which may be derived from the lock-in amplifier itself or from an external source, is applied to the apparatus. It may modulate the applied magnetic field, the radio frequency, the diffraction grating, or anything else that sweeps the signal. The modulated waveform, together with the noise, is phase-shifted if necessary and then delivered to the mixer that essentially multiplies it with modulation signal. The component of the detector waveform at the modulation frequency produces a dc level; all other frequency components produce an ac signal whose time average is zero. The output of the mixer goes to an RC filter that eliminates ac components in the signals. The lock-in amplifier can be viewed as a fixed-frequency Fourier analyzer whose bandwidth is effectively $1/RC$ and can therefore be very narrow. For example, $RC = 3$ sec produces a bandwidth of $\frac{1}{3}$ Hz, which at a typical modulation frequency of 300 Hz represents a filter with a Q of about 1000. This translates roughly into an enhancement of S/N of about 1000. Of course it is necessary to sweep at a speed that is slow enough to allow several time con-

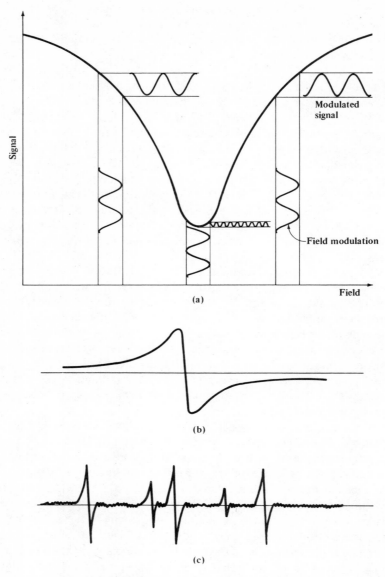

Figure 9.12 (a) Modulation of a signal produces sine waves of various amplitudes and phases. (b) A spectral line appears dispersion shaped after phase-sensitive detection. (c) A spectrum with several lines from a spectrometer with phase-sensitive detection.

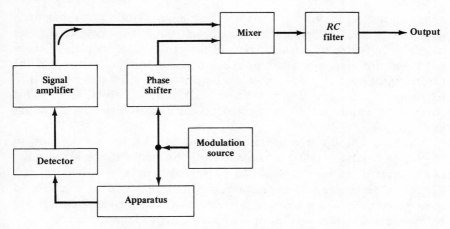

Figure 9.13 Block diagram of a lock-in detector.

stants per resonance peak. This may mean that sweeps last many minutes or even an hour.

It is possible to reduce the time constant, thereby increasing noise, but then make repetitive sweeps and signal average the results. At first thought it appears that this would produce a larger enhancement of S/N than would be achieved with either a signal averager or a lock-in amplifier alone, but this is not the case. A careful analysis of the problem shows that there is nothing to be gained by this technique in the improvement of S/N. However, instabilities and possible failures of some components in the apparatus may dictate circumstances in which this marriage of instruments is superior to using either of them alone.

D. Another major electronic instrument becoming more common in biological research laboratories is the small digital computer. In recent years, microprocessors have been replacing some minicomputers because they are even less expensive and more versatile. The result is that they can be dedicated to a single task and a laboratory of moderate means can afford to have several of them.

Applications of computers are limited only by the imagination of the users. As we have seen, computers can serve as effective signal averagers. They can also be programmed to simulate a lock-in amplifier or any other kind of data acquisition system. Their power lies not only in their ability to function like some other instrument but also to analyze data as it is coming in and/or to vary and control the experimental conditions. For example, if the signal amplitude becomes weak, they can raise the gain of the amplifiers. They can sweep the magnetic field slowly in the region of interest and rapidly through a region where no signal is expected. They can change samples, vary temperatures, detect problems and stop data accumulation when something goes wrong, analyze the data as it comes in, transmit the data to a larger computer for complicated analysis, and perform a large number of other tasks that are preprogrammed by the experimentor. This means that the experimentor needs to solve a problem only once (or a few times), give it to his computer, and then be free to think about scientific problems while the computer is performing the drudgery of repeated measurements and parameter variation.

We have already seen the beginning of the computer revolution in laboratories, but the best is yet to come. This is because of the rapid reduction of the price of microprocessors and associated peripheral devices. A new laboratory technology is presently being born that will have profound consequences for future research instruments. Computer-based devices will become as common as oscilloscopes in well-equipped laboratories.

Part 6: Suspensions and Solutions

The behavior of small particles or large molecules in solution or suspension can be studied to provide information about their size, mobility, or other properties. In these studies sample preparation is very important because the extraction of data relies heavily on the nature of the solvent and solute.

A. When an object is submersed in a fluid at rest, it is subject to a gravitational force from its weight and a buoyant force that results from the weight of the fluid. The net force, given by Eq. 3.2, is

$$F_{net} = (\rho_{object} - \rho_{fluid})Vg = mg\left(1 - \frac{\rho_{fluid}}{\rho_{object}}\right) \tag{9.24}$$

where V is the volume of the object (and the displaced fluid), g is the gravitational acceleration (g is a negative number), and ρ is the density. If the object is denser than the fluid, the net force on it will be negative (down) and it will sink. As it begins to move downward through the fluid, it will experience a frictional force proportional to its velocity v, so that the equation of motion becomes

$$m\frac{d^2y}{dt^2} = F_{net} - \beta v = F_{net} - \beta\frac{dy}{dt} \tag{9.25}$$

where β is the friction proportionality constant and y is the position of the object. We have seen in Part 1.2* that this equation of motion leads to a

---

*In this part, friction force was treated as being proportional to v^2 instead of v. The subject of choosing the quadratic or linear dependence of fluid friction is discussed in Appendix C.

terminal velocity when the object has reached a speed

$$v_t = \frac{F_{\text{net}}}{\beta} \tag{9.26}$$

and the acceleration is zero. We can solve Eq. (9.25) and find

$$y = v_t[t + \tau(e^{-t/\tau} - 1)] \tag{9.27}$$

where $\tau = m/\beta$ and the initial conditions require the object to be released from rest at $y = 0$. It is clear from Eq. 9.27 that after a long time the velocity dy/dt is simply v_t and the distance the object has moved is essentially $v_t t$.

We can evaluate Eq. 9.26 for the special case of a spherical object. The fluid friction force is given by Stokes' law as

$$F = -6\pi\eta r \frac{dy}{dt} = -\beta v \tag{9.28}$$

where η is the viscosity of the fluid and r is the radius of the object. For a micron size particle whose density is 10% greater than that of water we find $v_t \simeq 2 \times 10^{-5}$ cm/sec. If we wish to separate such particles from solution by allowing them to settle in a 10-cm tall test tube, it would take about 1 week.

In order to speed up the settling process, the suspension can be placed in a centrifuge that spins it in a horizontal plane (vertical axis) very rapidly. If we view the particle from a carefully chosen frame of reference which is fixed to the spinning test tube and rotating with it, the argument which leads to Eq. 3.2 can be applied. The force on an element of fluid of mass m_f rotating with the test tube is still $m_f g = \rho_f V g$ in the vertical direction but is $m_f \omega^2 r$ in the horizontal direction. This force is necessary to make the fluid rotate. Here, ω is the angular frequency of rotation and r is the distance to the rotation axis. The ratio of the horizontal-to-vertical forces is

$$\frac{m_f \omega^2 r}{\rho_f g V} = \frac{\omega^2 r}{g} \tag{9.29}$$

which is very much larger than unity for a typical centrifuge. We shall therefore ignore the vertical force and imagine the test tube to be tipped so that it is nearly horizontal. The rotational force is also horizontal and serves to keep the liquid in the test tube.

If the element of fluid were replaced by an object of density ρ_{object}, the object would be subject to the same buoyant force but it would require a force $\rho_{\text{object}} V \omega^2 r$ in order to keep it rotating with the fluid. If its density were higher than that of the fluid, it would move out to larger r (toward the bottom of the test tube) because of the imbalance of forces. The net force on the object is

$$F_{\text{net}} = (\rho_{\text{object}} - \rho_{\text{fluid}}) V \omega^2 r \tag{9.30}$$

which is similar to Eq. 9.24. We can write the equation of motion for a particle and solve it as before and the result is the same as replacing g by the centripetal acceleration $\omega^2 r$.

With a high-speed motor (say 10,000 rpm) and a rotor of 20-cm radius, the terminal velocity of a particle can be increased by a factor of 2×10^4 and the settling time for the suspension discussed above can be reduced from a week to half a minute. A modern ultracentrifuge achieves $\omega^2 r = 300{,}000g$, which enables the settling of extremely small particles whose density is very close to that of the solvent.

B. Careful observation of the light scattered by a solution or suspension of particles can provide a great deal of information. In order to understand the methods of light-scattering spectroscopy we must first discuss two completely independent phenomena called the *Doppler effect* and *beats*.

The Doppler effect produces a shift in the frequency of a wave scattered or reflected by a moving object. It is easiest to understand if we consider the process of reflection or scattering as if it were two separate steps: absorption followed by emission. The effect can then be discussed by studying the emission of a wave from a moving source.

It is important to remember that the speed of a wave in a medium (or of light in a vacuum) does not depend on the motion of the source. As long as the moving source has a velocity less than the wave speed, long after the wave has left the source, its speed depends only on the properties of the wave propagation. We shall restrict ourselves to cases where the source velocity is slow compared with the wave velocity and ask about waves detected from an approaching source. Figure 9.14(a) shows a source of waves of wavelength λ and speed v moving at constant velocity u toward a detector a distance d away. After a time interval Δt the source has moved closer to the detector by $u\Delta t$ and has emitted $f\Delta t$ new waves, where f is the frequency. It is clear that the number of waves in the space between the source and detector has been reduced by $u\Delta t/\lambda$ simply because the space between the source and detector has been reduced by the distance $u\Delta t$. The number of waves that have impinged on the detector during Δt is therefore $u\Delta t/\lambda + f\Delta t$, which means that the apparent frequency (cycles per/second) is

$$f' = f\left(1 + \frac{u}{v}\right) \tag{9.31}$$

since $v = f\lambda$.

This frequency shift also requires a change in the wavelength of sound or light emitted from a moving source, as illustrated in Figure 9.14(b). On the left we see circular waves of fixed length λ emitted in all directions from a stationary source. On the right we see that during the time a moving source has traveled a distance $u\Delta t$ the original waves have moved $v\Delta t$ and the source is no longer at the center of the circular wave pattern. The waves are com-

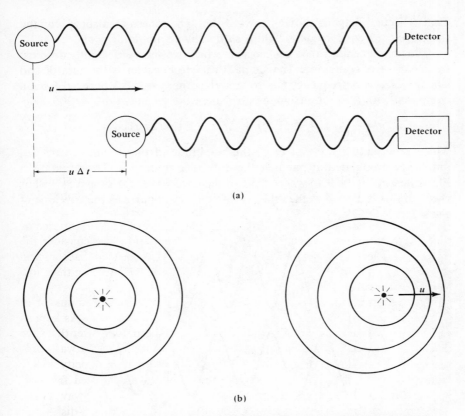

Figure 9.14 Waves emitted from a moving source. (a) The waves impinging on a detector. (b) The two-dimensional pattern.

pressed in the direction of travel of the source and stretched in the opposite
direction. Since their propagation speed is the same in all directions, and
since $v = f\lambda$ must be satisfied, the frequency is increased in the direction of
travel of the source, decreased in the opposite direction, and unchanged if
the source moves perpendicular to the direction of observation.

The scattering of light by a moving object has to be considered carefully.
If the object is moving at a speed u toward the source, similar arguments show
that it sees the frequency $f[1 + (u/v)]$ and therefore re-emits that frequency.
If the detector receives light that is scattered back toward the source, the
frequency it sees is higher than the emitted frequency by the same amount,
and (for the case of u very small compared with v) the Doppler shift is twice
as large; that is, the detected frequency is $f[1 + (2u/v)]$. Another way to
think about this is to consider that the detected light is coming from a virtual
source produced by mirror reflection of the source by the moving scatterer.

That reflected image is moving toward the light source at a speed $2u$, and therefore the frequency of its light is shifted by an amount $(2u/v)f$.

The phenomenon called *beats* occurs when two waves of slightly different frequencies are combined. The simplest case to consider is the sum of two sine waves, one of frequency $f - \Delta f$ and the other $f + \Delta f$, where Δf is small compared with f. We use trigonometric identities to find

$$\sin 2\pi(f - \Delta f)t + \sin 2\pi(f + \Delta f)t = 2 \sin (2\pi ft) \cos (2\pi \Delta ft) \quad (9.32)$$

which represents a wave of twice the amplitude of the original waves but amplitude modulated at one-half the difference frequency. The situation is illustrated in Figure 9.15 where the two separate waves are shown as well as their sum. It is clear that the difference frequency appears as the envelope of the sum.

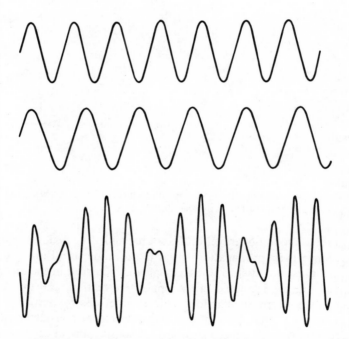

Figure 9.15 Two sine waves of slightly different frequency and their sum showing beats.

Of course, a Fourier analysis of the sum would show only the two original frequencies because those are all we started with. If some nonlinear device were to process the signal, however, the result would be the appearance of a frequency component at the difference frequency. Musicians are familiar with the presence of these beats and in fact make use of them to achieve precise tuning of their instruments. When the beat frequency between two closely spaced notes disappears, the notes have the same frequency. In this case the source of the nonlinearity is either the human ear or some parts of the musical instrument.

In 1965, George Benedek performed an experiment that made use of Doppler shifts and beats, which he called *light-beating spectroscopy*. He knew that small particles in suspension or solution moved around with velocities of a few centimeters per second as a result of Brownian motion. The Doppler shift of light scattered from these particles should therefore be $\Delta f = (u/c)f \cong 10^5$ Hz. No spectrometer could distinguish such a small frequency shift, but Benedek combined the Doppler shifted scattered light with some of the unshifted light reflected from the stationary container walls, mixed them at the photosensitive surface of a photomultiplier (a nonlinear device), and detected beats. (This technique of self-mixing or light-beating spectroscopy is possible only because of the very pure spectral properties of laser light.) There were beats over a large range of frequencies corresponding to the various speeds and directions of the scattering particles [recall from Figure 9.14(b) that the Doppler shift is directionally dependent]. Benedek was able to use the observed beat spectrum to study the Brownian motion (rapid thermal vibrations) of the suspended particles because their velocities were transformed into frequencies that were easily displayed by a spectrum analyzer resulting in a measure of the velocity distribution.

The random motion of particles in suspension or solution is characterized by an average velocity of zero, but by an RMS velocity that depends on the mass and size of the scatterers. The technique of light scattering can therefore be used to determine these parameters for various preparations of proteins, DNA, etc. Furthermore, it is possible to observe the time dependence of size and mass by repeating the experiment periodically and, therefore, to follow the progress of synthesis, decomposition, polymerization, etc., and to detect the effects of temperature, catalysts, drugs, etc., on these processes. Light scattering is a very powerful tool for providing a specialized kind of detailed information about many processes of interest.

The flow of blood offers interesting application for the technique of light scattering because the velocities of the moving blood cells can be measured. Consider the case of laminar flow discussed in Chapter 3, where the velocity distribution of the fluid and, therefore, of the suspended cells in it is given by Eq. 3.22. We calculate the thickness dr of a cylindrical shell needed to contain fluid with velocity in the range from u to $u + du$ by differentiating Eq. 3.22:

$$dr = \frac{2\eta L}{r\Delta P}\, du. \tag{9.33}$$

The volume of this shell, and the number of scattering centers contained in it, is proportional to

$$dA = 2\pi r\, dr = \frac{4\pi\eta L}{\Delta P}\, du. \tag{9.34}$$

We see that the number of scatterers with velocity between u and $u + du$ is independent of r and therefore independent of u. The amount of scattered light with frequency shift $(u/v)f$ is therefore a constant up to the maximum value of u. This maximum value is $u_{max} = \Delta P R^2/4\eta L$ and is related to the volume flow rate dV/dt by $dV/dt = (u_{max}/2)\pi R^2$ (see Eq. 3.24). We would therefore expect the spectrum of scattered light from blood flowing in accordance with Poiseuille's law to exhibit a region of constant intensity up to a certain maximum frequency and then drop to zero beyond that frequency. Feke and Riva have measured the flow velocity of blood in human retinal arteries and veins with sufficient precision to see the pulsatile motion arising from the heartbeat. Figure 9.16 shows the results of a single measurement. Its flat beginning followed by a rapid drop-off suggests that the flow is indeed in accordance with Poiseuille's law. The magnitude of the velocity can be used to determine the total circulation in the retina and therefore to diagnose circulatory disorders.

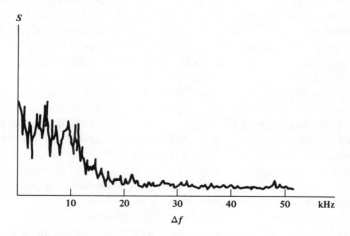

Figure 9.16 Doppler shifted frequency spectrum from flowing blood in the human retina.

The applications of light scattering are much broader than those discussed here. The method has an unpredictable potential for use in noninvasive diagnostics as well as medical research. There are many variations and combinations with other techniques that can extend its utility to many new areas of study.

Part 7: Ultrasound

In Chapter 7 we learned that the human ear is sensitive to frequencies of sound between 20 and 20,000 vibrations per/second (one vibration per/second is called one hertz, abbreviated Hz). We also learned that propagating sound waves are described as one set of normal modes of vibration of a complicated medium. However, the solutions to the equation of motion that give sound waves as normal modes are not restricted to frequencies within the range of human hearing; it is quite possible to produce vibrations at frequencies considerably higher and lower than those that are audible. The physiological effects of vibrations at frequencies less than 20 Hz are significant, but we shall not study them here. *Ultrasound* is the name given to vibrations with frequencies higher than the range of human hearing and typically has a spectral range from 20 kHz to 100 MHz. Modern ultrasonic techniques consist of sending a short pulse of high-frequency sound into the body and recording its reflections from the boundaries between organs.

The frequencies of ultrasound in modern medical diagnostics are restricted for practical reasons to the range between about 1 and 20 MHz, although there is certainly some limited amount of work outside of this range. Since the speed of sound in most human tissue is very nearly the same as its speed in water, about 1500 m/sec, the wavelengths used are generally in the range from 0.3 mm to 0.75 mm (only a few hundredths of an inch of an inch corresponding to 2–5 MHz). Generally the longer waves (lower frequencies) penetrate more deeply but provide less resolution of detail than the short waves.

Ultrasonic waves, like all other waves, exhibit the property of diffraction that limits the precision to which they can be focused or aimed. Equation 8.16 gives the minimum half angle of a beam. In order for an ultrasonic beam to spread by less than a centimeter or two after it has traveled across a human torso (about 30 cm) the source of the waves must be about 10 or 20 times larger than the wavelength or about 1 cm in diameter. Most commercial ultrasonic devices use the same probe for transmitting the ultrasonic wave into the patient and for recording the echoes of it, and their size is therefore chosen so that the beam spread is comparable to their diameter. For 0.5-mm waves, a 1-cm diameter source produces a beam with a spread of about $\frac{1}{20}$ radian (3°) resulting in a beam spread of 1 cm after traveling a distance of 20 cm. Diffraction of the returning beam simply requires a factor of 2.

A typical ultrasonic diagnostic machine is equipped with a set of interchangeable probes capable of operating at various frequencies. The active part of these probes is a special piece of material that is *piezoelectric*. Such materials, usually crystals, expand or contract when a voltage is applied to them and generate a voltage when they are mechanically distorted. They make excellent transducers for ultrasonic studies because they can be used both to generate the sound wave by the application of voltage and to detect the reflected sound wave (echo) by observing the voltage they put out as a result of the vibration.

The probes are connected to an instrument that causes them to emit short pulses of ultrasound and then record the returning echoes. After the pulse has been transmitted into the body, the echoes reflected from the boundaries between different kinds of tissue and organs are detected by the same probe and processed by various circuits for display and recording.

The length of the sound pulse is determined by practical considerations. If it is too long, the transducer will still be emitting when the first echoes return and they will not be detected. If it is too short, there will not be enough energy in the echo to detect easily. Furthermore, it can never be shorter than a few cycles of the waves because its frequency would then be undefined. At frequencies of a few megahertz, the shortest pulse that can be detected by the transducer in the probe is therefore about a microsecond long, which corresponds to a 1.5-mm length in the body. The longest pulse that would be sensible to use is given by $2d/v$ where d is the distance to the nearest reflecting surface of interest and $v = 1500$ m/sec. This is typically 15 to 20 μsec.

We have seen that in order to achieve a resolution of about 1 mm, diffraction requires wavelengths of a fraction of a millimeter resulting in frequencies in the megahertz range. In order to avoid echoes overlapping the incoming sound wave, the pulse must be only a few millimeters long, which corresponds to several cycles, and must therefore have a duration of a few microseconds. Since the last echoes of interest return to the probe after $2d'/v$ sec (d' is the distance to the furthest point of interest), the next pulse can be sent out after this time, which is about 500 μsec. Therefore the pulses can be repeated about 2000 times per second. These considerations define the operation of most typical ultrasonic devices.

The simplest display of a set of echoes is shown in the top part of Figure 9.17(a). Every 500 μsec a pulse is emitted and it is followed by several echoes over the next few hundred microseconds. The pulse heights represent the amplitude of the echoes and the horizontal position indicates the time. It is sometimes convenient to display only spots at the time when the echoes occur and to indicate the strength of the echoes by the brightness of the spot. Such a display is illustrated in the lower part of Figure 9.17(a).

If the probe is moved to a new position or aimed in a different direction, the time between echoes will be changed. By moving the probe around the body, it is possible to obtain a display of dots that outline the reflecting surface as shown in Figure 9.17(b). This picture provides only a two-dimensional outline, but three-dimensional information can be obtained by moving and aiming the probe properly. Of course, presentation of the three-dimensional information is not a trivial problem. Sometimes sophisticated computer

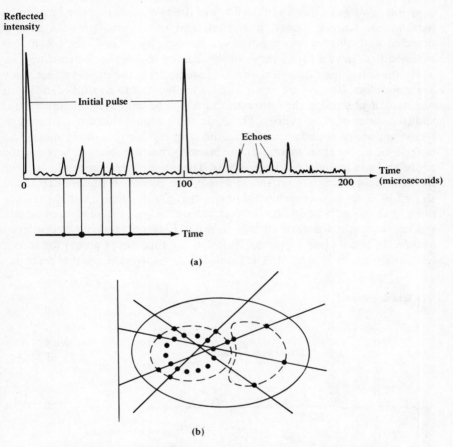

Figure 9.17 Ultrasonic echoes. (a) The echoes received plotted against time in a single dimension. (b) The results of carefully displaying a set of echoes taken from different directions. The internal organs are well defined.

graphics procedures are required to process and present an intelligible and useful three-dimensional picture.

The techniques described above are used to provide "pictures" of the insides of a patient. For example, such a study can determine the location and size of an aneurism in order to provide guidance for a surgeon. It is possible to observe a fetus, determine the presence of twins (or triplets, etc.), measure the size and position of the baby, determine the location of the placenta, and provide information about a number of things of interest to an obstetrician. Cancerous tumors are often observable in the lungs, breasts, or abdominal organs where their size and rate of growth can be determined over a period of time. In this way the effects of drugs or radiation therapy can be ascertained. Kidney stones, gallstones, and other hard growths can be detected and observed with ultrasonic devices. In general, the techniques outlined here have a broad range of application in diagnostic medicine.

In the foregoing discussion we have assumed that the organs under study are motionless. If a moving organ, such as the heart, is to be studied, different methods of presenting the information need to be employed. Figure 9.18(a) shows a series of chest echoes. There are some echoes that come from relatively stationary boundaries such as the lung surface (breathing moves the outer wall of the chest as well as the inner so that the distance between the probe and the boundary is unchanged). There are other echoes which come from surfaces which are moving. Some of the echoes in Figure 9.18(a) are shifted by 5 to 10 μsec, which indicates that the boundaries that produce them have moved 8 to 15 mm between the traces. A large number of closely spaced, sequential traces of the type shown in Figure 9.18(a) can be displayed as a series of dots [see Figure 9.17(a)] that are then swept across the screen while the probe is held fixed. The result is a set of wiggly lines that trace out

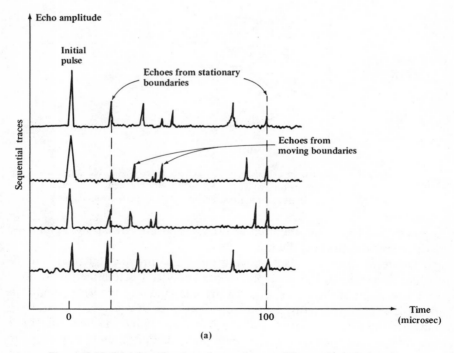

(a)

Figure 9.18 A series of pulse echoes taken at different times during the heart cycle is shown in (a). Some of the peaks have moved and others have not. (b) The temporal display of the echocardiogram.

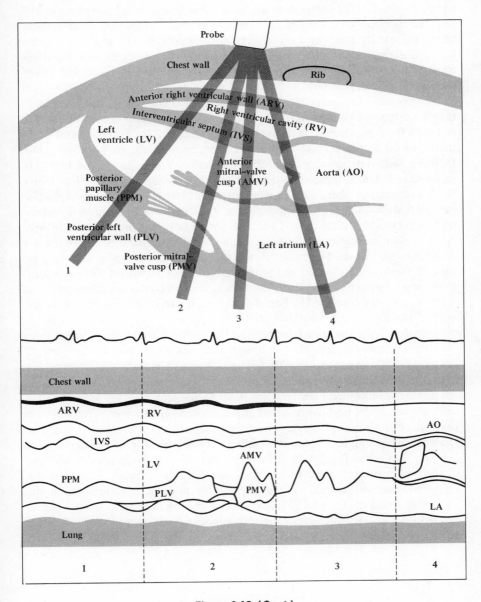

Figure 9.18 (*Cont.*)

the path of the various reflecting boundaries along the line of the ultrasonic beam. Figure 9.18(b) shows such an "echocardiogram." One can observe not only the motion of the walls of the heart but also that of the individual leaflets of the valves! By studying a large number of normal cases, cardiologists can learn to diagnose pathological cases just as they can with ECG's. Furthermore, it is possible to pinpoint defects for corrective surgery as well as study the results of such surgery.

When sound waves (or any waves) are reflected from a moving boundary, they undergo a Doppler shift (Eq. 9.31). For the case of ultrasonic waves reflected from moving parts of the heart, u is about 2 to 20 cm/sec and Δf is therefore 60 to 600 Hz. The difference frequency produced by mixing the Doppler shifted echo with the original ultrasonic frequency is audible. It is possible to listen to these sounds and hear only the motion of the heart surfaces without the heart sounds usually encountered in phonocardiography (see Part 4.3). Doppler shift techniques in ultrasonics are presently under rapid development and should become a standard tool in the near future.

It is often convenient to use a continuous wave source (abbreviated CW) for ultrasonic studies. For example, if a CW sound is directed at an artery or vein, the reflected wave will be Doppler shifted by an amount corresponding to the velocity of the blood flow. Ultrasonic velocimetry is complementary to laser velocimetry discussed in Part 9.4 because it can be used deep inside the body. Both have the advantage that the flow pattern can be studied without the insertion of a catheter (see Part 3.4), which always involves some risk.

Ultrasonic holography is a technique that has great promise for medical use but has the disadvantage that it is difficult to extract the information from the ultrasonic hologram and present it in a usable form. The methods are analogous to optical holography. A large diameter CW beam of sound is directed at a body, and the reflected and scattered waves are interfered with a portion of the original beam. The two-dimensional interference pattern is recorded by letting it strike and distort a very flexible, highly reflective sheet (often a thin metal foil). A laser beam is also reflected from the foil and the distortions produce nonuniform reflection of the laser beam. The resulting distorted laser reflection is photographed for later observation. Often the subject to be studied is placed in a tank of water and the reflective surface is floated on top. This provides a smooth support for it as well as good ultrasonic conduction (through the water) from the subject to the foil.

APPENDIX J
PHOTODETECTORS

The purpose of most photodetectors is to produce an electrical signal of magnitude proportional to the intensity of an incident beam of light. This signal can be measured with a meter, displayed on an oscilloscope, or fed to some data acquisition device for recording or analysis. Laboratory measurment devices do not convert light energy into electrical energy in the sense that a device to utilize solar energy might be employed, but instead are intended to provide accurate measurement of the incident light level and variations of it.

One of the simplest photodetectors is the vacuum diode, illustrated in Figure J.1(a). Light is incident on a curved piece of metal called the *photocathode*. Electrons of charge e are ejected from the cathode (the photoelectric effect) when the frequency f of the light is high enough to satisfy $hf > e\phi$ where h is Planck's constant and ϕ is a property of the metal called the *work function*. For most metals, ϕ is so large that only high-frequency light (ultraviolet) will cause electron ejection and therefore photocathodes are usually coated with special materials chosen for their low work functions and high photoelectric efficiencies.

The battery (typically 50 to 100 volts) in the circuit in Figure J.1(a) produces a voltage that accelerates the photoelectrons toward the anode where they cause a current to flow in the circuit resulting in a voltage across the resistor. This voltage is measured or recorded. Some photodiodes contain gas that is ionized by the accelerated photoelectrons resulting in more charges at the anode for each photoelectron and thus a larger signal.

The most common photodetector in general laboratory use is a *photomultiplier* tube (PMT), which is also based on the photoelectric effect. Figure J.1(b) shows that the PMT also has a photocathode that ejects electrons when light is incident. The vacuum envelope of the PMT contains a series of *dynodes*, that are supplied with a sequence of accelerating voltages (usually all from the same power supply of typically 1000 volts) in order to accelerate the electrons from one to the next. Typically a single photoelectron from the cathode is accelerated through about 100 volts to the first dynode where its impact releases several new electrons. Each of these is accelerated to the second dynode by another 100 volts, and their impacts each release several new electrons. Typical PMT's have an array of 10 dynodes resulting in about 10^6 electrons for each photoelectron. This shower of electrons strikes the anode during an interval of a few nanoseconds resulting in a pulse of 0.1 mA or 5 mV into 50 ohms.

Notice that in the circuit of Figure J.1(b) the last dynode is kept about

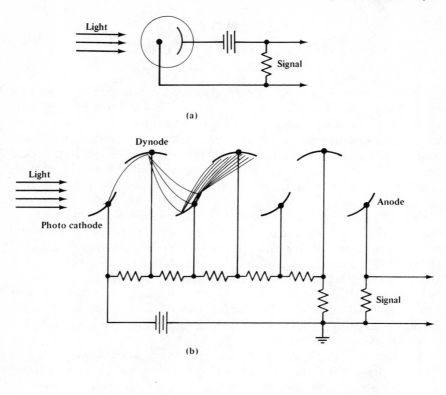

(a)

(b)

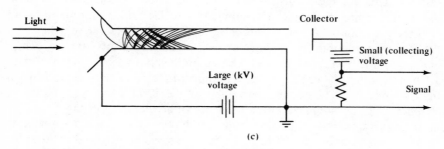

(c)

Figure J.1 A photodiode in a typical circuit is shown in (a). A photo-multiplier circuit that illustrates how the electron shower forms is shown in (b). Usually the dynodes are oriented in a circle. The operation of a channeltron is shown in (c). This continuous dynode device employs a collector plate for the electron shower.

100 volts below ground potential by the final resistor in the chain, so that electrons are accelerated to the anode even though it has no connection to the power supply. The signal can be processed as before, but it is also pos-

sible to amplify the pulses and then count them electronically. A system to do this is loosely called a *photon counter*, but it is actually a photoelectron counter.

The most recent advance in the technology of photoelectric detectors is the continuous dynode device called a *channeltron*, illustrated in Figure J.1(c). A very narrow tube is internally coated with a resistive material and subject to a high voltage. A cone-shaped photocathode is mounted near the end of the tube, and the photoelectrons are accelerated into the tube where they strike the walls and release more electrons. The result is also a shower of electrons that produces an easily handled current pulse like that of the PMT. An array of tubes (without cones) has some spatial discrimination and can be used as an image detector or enhancer. It is called a *channel plate*.

In addition to the vacuum tube devices described above there is a variety of solid state detectors. The most common ones employ a semiconductor appropriately doped so that its conductivity is minimal. When light is incident on the material, electron-hole pairs are created and the material conducts. A bias voltage [see Figure J.1(a)] sweeps out these mobile charges and therefore draws a current that constitutes the signal. Some of these are also photovoltaic: they produce a voltage without the need for a power supply and are therefore much more convenient to use. Often a small voltage amplifier is built into the same package with the detector.

Some photodetectors produce a signal derived from the total energy delivered by the incident light. The light is absorbed by a piece of material and its energy is converted to heat. A thermometer measures the temperature change and produces an electrical signal. These *thermopiles* are relatively slow devices but can be accurately calibrated and have broad spectral range. They are especially useful in the infrared where the light does not have a high enough frequency to release a photoelectron or produce an electron-hole pair.

The principle problems that limit various photodetectors are their spectral response, efficiency, dynamic range, noise, and temporal response. There is no ideal device for all applications, and a discussion of the various properties would take too much space. In general, PMT's are fast, linear, and very sensitive but have a limited spectral range and are sometimes noisy. Also, they require kilovolt power supplies. On the other hand, solid state devices have a broader spectral range, somewhat less sensitivity, and lower efficency but operate from lower voltages. Thermopiles respond from the far ultraviolet to the far infrared, but are slow and not very sensitive. They are often chosen for standards calibration.

Questions and Problems for Chapter 9

1. Why is spectroscopy by Michelson interferometer called Fourier transform spectroscopy?

2. How far would a mirror have to be moved in a Fourier transform spectrometer in order to resolve the sodium D-lines (5890 and 5896Å)?

3. From Figures 9.4 and 9.5 it is clear that the binding energy of a molecule is typically a few times the vibrational energy. Since vibrational frequencies are typically 10^{13} Hz, calculate a typical molecular binding energy.

4. The separation between the atoms in the oxygen molecule is about 1 Å, and each of them has a mass of about 16 times the proton mass. Calculate the moment of inertia of the molecule and verify Eq. 9.6.

5. Perform the suggested algebra and derive Eq. 9.18.

6. Suppose a gamma ray detector responds to 1% of the radioactive decays of I^{131} in a patient's thyroid gland. How many curies of radiation must be deposited there in order to detect its presence in one second? How many curies to detect a 10% change in the level of radiation in one second? (A curie produces 10^{10} radioactive decays in one second.)

7. When light is reflected from a moving object, it is often convenient to consider the reflected light as if it came from a moving, virtual source which is the image formed by the reflector. Show that, if the reflector is moving at velocity v, the moving virtual source is moving at velocity $2v$.

8. Show that Eq. 9.27 is a solution to Eq. 9.25 and that it satisfies the initial condition $y = 0$ at $t = 0$. Differentiate it and plot velocity vs. time.

Bibliography

1. E. ACKERMAN, *Biophysical Science* (Prentice-Hall, Inc., Englewood Cliffs, N.J., 1962). See especially Chapters 26–31. (An excellent treatment of a number of topics in biophysics.)

2. B. CHU, *Laser Light Scattering* (Academic Press Inc., New York 1974). (A research-level text book.)

3. CLARK, LUNACEK, and BENEDEK, *Am. J. Phys.* **38**, 575 (1970).

4. G. DEVEY and P. WELLS, "Ultrasound in Medical Diagnostics," *Scientific American* (May 1978).

5. T. EVERHART and T. HAYES, "Scanning Electron Microscope," *Scientific American* (January 1972).

6. G. FEKE and C. RIVA, *J. Opt. Soc. Am.*, **68**, 526 (1978).

7. D. FITZGERALD, W. BRAY, C. FORTESCUE-WEBB, and C. DONNELLY, J. Irish College of Physicians and Surgeons, 5, 11, (1975). (A discussion of ultrasound for aneurism detection.)

8. N. FORD and G. BENEDEK, *Phys. Rev. Lett.* **15**, 649 (1965).

9. D. HALLIDAY and R. RESNICK, *Physics* (John Wiley and Sons, New York, 1962). (Any other standard introductory physics text with a discussion of microscopes may also be used.)

10. G. HERZBERG, *Atomic Spectra and Atomic Structure* (Dover Publications, Inc., New York, 1945). (An introduction to structure and spectroscopy.)

11. R. HOBBIE, *Intermediate Physics for Medicine and Biology* (John Wiley & Sons, New York, 1978). (An excellent treatment of many topics of great interest. Highly recommended.)

12. P. LEBOW, F. RAAB, H. MET CALF, *Measurement of g-factors by Quantum Beats in the OH Free Radical*, Physical Review Letters *42*, 85 (1979). (A discussion of molecular spectroscopy with references to papers on laser development.)

13. Diffraction Grating Handbook, Staff of the Research Laboratory, Bausch & Lomb, Inc., Rochester, N.Y. 1970.

INDEX